COMPUTERS IN THE LIFE SCIENCES

COMPUTERS IN THE LIFE SCIENCES

APPLICATIONS IN RESEARCH AND EDUCATION

EDITED BY R. LEWIS

CROOM HELM LONDON

Croom Helm Ltd, 2-10 St John's Road, London SW11

British Library Cataloguing in Publication Data

Computers in the life sciences.
1. Biological research – Data processing
2. Biology – Study and teaching – Data processing
I. Lewis, R
574'.072 QH324.2

ISBN 0-85664-863-9

Printed in Great Britain by
Biddles Ltd, Guildford, Surrey

CONTENTS

Editor's Preface

1. Background and Perspective 9
   1.1 Introduction 9
   1.2 Quantitative Models in Research and Education 9

2. Quantitative Methods in Science 12
   2.1 Science, Science Education and Mathematical Models *R. Lewis* 12
   2.2 Quantitative Treatment of Medical Data *G. Franglen* 16
   2.3 Mathematics in Scientific Research and Education *B.V. Gnedenko* 23
   2.4 Some Difficulties 25

3. Computers in Life Science Research 26
   3.1 A Review of Applications 26
   3.2 Mini Computers Interfaced to Gamma Scintillation Cameras in Nuclear Medicine: The Role and Achievements in the Development of Nuclear Cardiology *V.B. Bosnjaković* 27
   3.3 The Analysis of 'Thermal' Oscillations in Heart Rate Variability *R.I. Kitney* 41

4. Modelling in the Teaching of Mathematics 56
   4.1 Introduction 56
   4.2 Some Uses of Population Models in the Teaching of Mathematics *B.W. Conolly and F. Oliveira-Pinto* 56
   4.3 The Place of the Computer in Teaching Applied Mathematics *R.D. Harding* 67

5. Computers in Teaching Biomedical Science 77
   5.1 Introduction 77
   5.2 Teaching Biological Systems and Signal Analysis to Undergraduate Physiologists *R.I. Kitney* 77
   5.3 Transparent 'Black Boxes' *A.W. Wood* 85
   5.4 Confronting Students with Mathematics 93

6. Man-Computer Communication 94

6.1 Introduction 94
6.2 Towards a New Computer Language *R.D. Harding* 94
6.3 The Delay Concept in Computer Simulation Languages and the Role of These Languages in Modelling and Education *F. Oliveira-Pinto* 98
6.4 An Approach to the Design of Problem-oriented Languages, with an Approach to Neurophysiological Experiments *N. Parezanović, B. Janković and Z. Damjanović* 104
6.5 A Hierarchy of Interactions 120

Notes on Participants 122

Index 123

EDITOR'S PREFACE

This book arises from an Anglo-Yugoslav symposium on mathematical methods and computer models in research and education, held under the patronage of the British Council and the Yugoslav Union of Science Societies, at Herceg-Novi, Yugoslavia, in November 1977.

A symposium which seeks to break new ground is best planned with just a light framework with which to catalyse discussion, and with a dynamic structure, free to encompass the widest possible arguments. However, this does leave the editor of its proceedings with the daunting task of presenting the contributions, both formal and informal, in a way which the reader will find coherent and illuminating.

This book contains papers in sections of the various chapters, which are attributable to one or more authors who are identified along with their contribution. Other sections, without named contributors, are the responsibility of the editor, although often the ideas contained in them stemmed from discussion during the symposium and the working groups which met long into the nights. My thanks go to all the symposium members listed, who have encouraged me in my task and in particular to my colleagues Zvonko Damjanović and Dick Kitney for their advice and support.

R. Lewis
July 1978

1 BACKGROUND AND PERSPECTIVE

1.1 Introduction

Today's students are tomorrow's research scientists. It is not possible to conceive of a scientist in the 1980s who does not have a broad perspective of perhaps the most important tool of science, the computer. Further, many of the most challenging problems and often the most rewarding aspects of science are those which may be described as multi-disciplinary. The contributions made by physics to medicine and by mathematics to ecology, are but two of many areas of science which will mould the shape of mankind in the next century.

It was, however, with much more limited horizons that teachers and research workers from Yugoslavia and Britain agreed to collaborate on aspects of their work which may be termed multi-disciplinary. The use of mathematics and computing in this context was considered to be one appropriate theme and so, after exploratory discussions, a symposium on mathematical methods and computer models in research and education was held under the patrongage of the British Council and the Yugoslav Union of Science Societies. This first symposium was held in Herceg Novi under a general collaborative agreement between the Centre for Multi-disciplinary Studies in the University of Belgrade and Chelsea College in the University of London to pursue key issues in science education and research.

This publication attempts to reflect the views of the participants at this symposium. As will become clear, their expertise lay in various fields but all carried a common purpose, that of increasing understanding between workers from different disciplines.

1.2 Quantitative Models in Research and Education

The early chapters describe the positions, and the occasional difficulties, of both scientists and mathematicians in undertaking research and the related education and training needed. Later some tentative paths forward are proposed. However, some issues relating to the development and use of models which were identified in the discussions will be presented first in order to provide a framework for later chapters.

When developing a computer model for the purposes of teaching, it is usually desirable that the model comes from previous research work. The main reason is that students should feel that the model being used relates directly to the real situation and that their investigations in some way reflect the original research.

Extending the concept of models derived from research for use in teaching, it is important to design the model so that it can be used at several levels of tuition. Such an objective requires that at low levels certain areas of the model should effectively be black boxes, but also that these areas should be amenable to conversion to transparent boxes when required. Consequently, the model should normally be a direct representation of the process it is simulating. Also, variables within the transparent boxes should be capable of representation in a number of different forms, e.g. graphical, numerical or diagrammatical. These alternative methods of presentation will help the student to use the model in increasingly sophisticated ways, moving through diagrammatical to graphical presentation and finally by analysis of the numerical data to a consideration of the mathematics which underlies the model.

The development of the type of multi-level model outlined above will probably necessitate the development of a new type of computer language. There are a number of difficulties associated with the synthesis of such a language. A dilemma exists in the way the new language should reflect universal features of a language like FORTRAN or the more specific features of a language like GPSS. One possible way of resolving the dilemma might be to use a language based upon characters, as described in Chapter 5. A number of points should be borne in mind when developing the new language. While there is a need for the operator blocks to be based on the equations 'found' in the process, a better approach would be to identify, where possible, common subprocesses and design operator blocks to solve the equations which underlie these subprocesses. The advantage of this approach is that it would reinforce for the student the common strategy which underlies many biological systems and break the mathematical description into appropriate biological subunits.

Another feature which must be taken into account in the development of a new simulation language is the different types of student who will use it. The language must be capable of allowing models to be built which are capable of rearrangement depending on the student audience. If, for example, the model is being used by medical students it is probably unnecessary and in fact undesirable for them to be able to access the part of the program which prints out the underlying equations or that part that even prints out data in tabulated form. On the other hand if the students are biophysicists there will be little point in their using the model if they could not access this information.

Whilst many simulations are continuous in nature, this is certainly not always the case in biology. For example, many scientists consider that the process of diffusion through membranes is a discrete, or more accurately, a queueing process. Related to this, it should be noted that biology frequently uses existing theory from other disciplines to describe biological processes. An example is the cable theory model for the propagation of an action potential along a nerve axon. While superficially attractive, such an approach means that the behaviour of the biological process being modelled is automatically constrained by the existing theory, with all its inherent assumptions and errors. To avoid such pitfalls, a better strategy would seem to be to return to a new direct mathematical description of the process.

In the design of a model it is also necessary to consider the degree of precision required. Mathematical models are frequently designed to provide an 'accuracy' far greater than that commensurate with the available experimental data. An error of this type can cause problems for the experimenter in discriminating between suitability of different types of mathematical models or indeed the degree of sophistication required in one type of model. In a multi-disciplinary study errors of this type are likely to cause serious misunderstandings between workers from different disciplines.

2 QUANTITATIVE METHODS IN SCIENCE

2.1 Science, Science Education and Mathematical Models

R. Lewis

Physics is often indistinguishable from mathematics as mathematics has often been invented to model physical systems, for example, calculus for the laws of motion. Once models come into common use, their use grows rapidly as fundamental laws, e.g. the inverse square law or exponential laws, are discovered to apply to more and more different systems. This is seen in the case of the atom, for which a model proposed was similar to that previously used to describe planetary motion. It is then possible to search for phenomena in one system corresponding to features in a second system at that time unrecognised in the first. In other words, one learns about a system by looking for correspondences with features of another. This is the view of the physicist.

Certain areas of science, however, do not have this tradition. To what extent do biologists think in this way? Does a membrane have to be like a coffee filter or can it be thought of in terms of a model which also accounts for movement under electrical potentials? If the latter is true how natural is it to see this as:

$$j_i = -\frac{1}{r_{ii}}\frac{d\bar{\mu}_i}{dx} = -\frac{1}{r_{ii}}\frac{d}{dx}(\mu_i^o + RT\ 1na_i + Z_iF\Psi + \bar{v}\,p).$$

How many other such models exist in biological systems and how much is learnt from the possible correspondences referred to earlier? It is said that much of biology concentrates on individual organisms each of which is different from others of the same species. At the same time, although individual behaviour may be different, certain patterns of collective behaviour can be identified. A close similarity exists here through the macroscopic (exponential) behaviour of radioactive substances and cell division, both processes, at the microscopic level, being apparently random in time. However, most physical systems have models which are deterministic and this cannot be said of many (any?) microscopic biological models and perhaps few macroscopic ones. Does this kind of difference between the two branches of science help us account for their different dependence on models?

Models have become more and more complex in recent years as it has been possible to build these into computer programs. In fact, many present-day models would be inconceivable if computers had not been invented. This implies that much theoretical and experimental work, now commonplace in science, would be impossible. Physical science made use of mechanical computational aids years before electronic computers were invented, e.g. Fourier analysers in the production of tide and astronomical tables. More recently biology has benefited from the use of computers in many ways, as in ecological models and pattern recognition.

Despite intrinsic resistance to change, mathematics and computing are making significant inroads into biological research, and the science is changing its nature in response. To the non-specialist this effect is seen through the debates arising from the use of models of the world. The study of various world models teaches us much about the use and misuse of computers. Quite often such controversies polarise views, and predictions from the same model can be interpreted as 'evidence' supporting quite opposite views. Many present-day biologists are of a pre-quantitative generation and so in attempting to introduce this approach in teaching we may be trying to do things too quickly. However, a biologist has recently expressed the view[1] that:

> In recent years an increasing emphasis has been placed in school biology on the study of the processes by which individuals and populations of living organisms operate. Consequently, less attention had been paid to the structures which are involved. This change has been accompanied by an experimental rather than a descriptive approach to the subject. The results of practical investigations are often in a numerical, rather than a pictorial form, although histograms and graphs may be drawn from the numbers.

The change in physics took place such a long time ago that we have no idea what growing pains it experienced although we do know that Galileo was considered a heretic for proposing a planetary system not centred around the earth. However, biology has not escaped controversy: Darwin's evolutionary theory contradicted the entrenched beliefs of the seven day creation of the earth. In fact the present controversy regarding inherited and environmental factors in intelligence is one such, this time involving political rather than religious dogma!

Naturally, curriculae and teaching methods lag some time behind the developments in the subject, so whilst physics education has followed the mathematical methods and computer models approach for some years, education in the biological sciences is barely reaching this stage. But surely it is inevitable that, since the science itself has gone along the path of increased use and eventual dependence on mathematical or at least model ideas, education in this field will follow the same lines.

The popularisation of computer models, for example those of the world referred to earlier, is affecting attitudes towards computing and models; at the same time, technology is bringing computation within the grasp of many previously remote from it through the commonplace use of hand calculators. The next generation of university students – having used calculators and probably micro computers in schools – may be rather more numerically aware than their predecessors. Succeeding generations will undoubtedly also be familiar with rather sophisticated information retrieval systems through their home telephones and television sets if not actually with computers inside their televisions!

So we are in an age where attitudes are changing radically and perhaps the present-day antipathy towards mathematics and computing from large portions of the population may be a passing phase. However, how do we begin to solve our present problems?

At a time when the volume of factual information is so large that it cannot be encompassed, even within narrow bounds, by individual human beings, it is increasingly important in all fields of knowledge to develop a methodology which is systematic. So it is seen that advances in any field, even those traditionally considered intuitive or 'unscientific', depend more and more on systematic development. This evolution in method is becoming an essential part of the way in which our students are educated.

If students are to be effectively educated for work in the 1980s their present university courses must provide an appropriate emphasis on computing, more specifically on computer applications in their discipline. The applications range from straight computation, through simulation to data analysis, automatic data capture and control. Details of courses and experience to be gained will vary, but amongst the topics to be included are likely to be:

> laboratory data analysis; real-time data reduction; experimental control; familiarity with and use of computer software packages; programming or program formulation; model building and simulation techniques; logic and circuit design and the use of micro processors.

Certain of these are clearly important for laboratory-based sciences whilst others will be emphasised in the social sciences. It should be noted that the inclusion of such topics in an undergraduate course is not intended to produce computer specialists, but to produce, say, biochemists, with a broad appreciation of the role and value of computing in biochemistry. The importance of such training was acknowledged some time ago in a United Kingdom report[2] which included the following statement:

> We think it important that young men and women entering upon their careers from university should accept computers and be able to understand them as the sophisticated tools they are, and the undergraduate stage, in our view, is the time to do this when what is learned is assimilated in a way not possible later. We may expect that in time every school child will receive some introduction to the use of computers, so that what is provided for university students will represent an increase in sophistication rather than something quite new. But meanwhile, because of their present novelty, we have with computers to educate *ab initio* at the university stage. We have no hesitation therefore, in advising that all undergraduates should be taught computing.

In developing courses for students in particular disciplines, a great deal of care is required. Much criticism has been made of many service courses, particularly in mathematics or statistics, as they have not been relevant or sufficiently motivating for students. Students are told that the techniques are important and must be learned; but this is not sufficient. A very close integration is needed to show clearly the immediate relevance of the 'ancillary' matter to the primary discipline.[3] To most students the use of computers in their discipline is at best remote and at worst irrelevant.

One way of beginning to overcome this problem is through the almost incidental use of computers in fairly traditional courses. It would appear that analysis of data from laboratory or field investigations naturally leads to the use of computers, but in fact it is fairly uncommon for student investigations to generate data of the volume or complexity to warrant computer reduction. (Beware of contrived situations: students usually see through them and hate being misled!) However, if the computer is used to assist directly in students' learning, then it can be incidental and yet immediately relevant. It is possible that Computer Assisted Learning (CAL) can play an important role in providing such experience for undergraduates and this will be considered considered more fully in Chapter 5.

References

1. M.E. Leveridge, *Computers in the Biology Curriculum* (Schools Council/Edward Arnold, London, 1978).
2. K. Berrill and B.H. Flowers, *Teaching Computing in Universities* (HMSO, London, 1970).
3. R. Lewis, 'Mathematics for Science – A Problem of Communication', *International Journal of Mathematics in Science and Technology,* vol.4, 1973.

2.2 Quantitative Treatment of Medical Data

G. Franglen

Physicians are expected nowadays to recognise that mathematics can be of value in some of their work and teaching. Many make a show of so doing; medical schools generally recognise that 'computers are useful. . .' or 'mathematics is a good thing. . .'. Too often, however, this obeisance to science and mathematics appears to be little more than one of following a fashionable attitude. This can be seen in many medical courses in the United Kingdom where instruction is divided precisely into two parts. The first two years are usually devoted to scientific studies essentially as pure disciplines with often little direct reference to the role they may play in medicine; this is followed by about three years of clinical practice where learning is based on an apprenticeship with little overt incorporation of the scientific approach taught over the first two or three years of the course. We have noticed that medical students, in general, do not easily relate for themselves the science and mathematics of the beginning of their course to their clinical studies. Most of them, after all, are wanting to be doctors and treat people, not to be mini-scientists, and they appear to forget very rapidly their pre-clinical studies as soon as they enter the wards and start dealing with patients. This is not only a waste of teaching time, it is also a waste of the student's motivation.

There is now, however, a growing acceptance that science, mathematics, and computers are of genuine value in medicine and, properly incorporated into a doctor's training, should improve medical skills. In a scientific and technological age we need doctors who regard scientific thought as a natural part of their background; it does not follow that, because one is scientific in dealing with disease, one will be a poor, inhuman clinician. Some planners of medical curricula are beginning to recognise the value of not splitting the course too clearly into two distinct parts, a scientific followed by a separate clinical section; there are now a few medical schools where teaching emphasises the relationship between science, mathematics and medicine throughout the course. It is hoped by this means to produce doctors who regard scientific disciplines as natural adjuncts to their clinical skills rather than as things which have to be applied when necessary, as, for example, to a paper to make it 'scientifically' acceptable for publication.

The traditional teaching methods of the clinician will continue to prevail for some time yet, however, and this is understandable. The traditional courses, after all, have consistently produced effective doctors without relying too much on pure science and mathematics. The clinician is often dealing with signs and symptoms in his patients which are not quantifiable in a mathematical sense; I have called this material 'soft data' in contrast to that which can be measured more or less precisely and which might be called 'hard data'. It is important never to forget that in many clinical situations much of the material on which the clinician is having to base his assessment is 'soft data', data which may not be amenable to precise analysis. Nevertheless, especially with the present need to train doctors to a high level of skill quickly and efficiently, it is important to be able to assess the value and nature of different types of clinical data for effective teaching, if for nothing else. If we do not make some attempt at this sort of assessment, communication about data may continue to be inadequate, as may be seen in the following personal example. My first medical research chief was an excellent diagnostician of the old school. On one occasion I was with him in a metabolic ward when he noticed for the first time a newly entered patient in a bed some twenty metres away. 'That woman has pellagra', he said. I protested: people did not get pellagra in Britain any more. We walked across to the patient; we looked at her more closely and at her clinical notes and, sure enough, she had pellagra. I was impressed, but I was not going to settle for just this demonstration of his skill; I wanted to know how he had come to his conclusion. After half an hour of questioning, the best I could get out of him was that this woman had reminded him of the look of a patient he had seen some 20 years ago who had also had pellagra; he was quite unable to tell me the characteristics of this 'look'.

This, of course, is an extreme example; many doctors use a mixture of 'hard' and 'soft' data in making diagnoses and in assessing the progress of their patients. The 'soft' data consists of items such as the general attitude of the patient as seen by the physician, the patient's account of himself; the 'hard' comprises of items such as measurements of pulse rate or laboratory examination of the blood. With regard to the latter, it should be noted that clinicians nowadays get much 'hard' data in the form of extensive chemical analyses of blood, serum and urine; this has arisen through the development of auto-analysis.

Laboratory data on patients is clearly a form of 'hard' data and, as such, is amenable to computer processing. This leads naturally to consideration of whether a reliable relationship can be shown to exist between this 'hard' data and the disease affecting the patient, and there have been a number of attempts at developing medical diagnostic programs, attempts, in a sense, at a partial model of the process of decision-making involved in the act of diagnosis. A difficulty which is sometimes overlooked is that one of the most important items on which such studies are based, the diagnosis itself, must be regarded in very many cases as a 'soft' datum. We can never be absolutely sure of the accuracy of every diagnosis, even when death has intervened and a pathologist has submitted his report. Nevertheless, such models are useful; they can help towards distinguishing between essential and non-significant data on which we base our decisions in this field, and this should improve the quality of clinical teaching. Moreover, the models occasionally show up relationships between items not previously associated which may prove to be of use in diagnosis.

As an illustration I will outline the development of a simple and useful computer program based on a study of laboratory chemical data as supplied to clinicians. It arose from the fact that auto-analysis has not been an entirely unmixed blessing; there are several difficulties which arise from its capacity to produce a very large number of analyses, with relative ease. For example: it is not always clear which analyses are of value in treating a particular patient since, to a significant extent, the auto-analytic apparatus has been put together for chemical rather than clinical convenience. It is easy to analyse sodium and potassium together, and so we analyse both; it does not follow that the clinical need to analyse both has been established.

The second difficulty is that chemical results (or 'parameters' as the clinician likes to term them) are usually expressed in scientific units. Thus, in a normal adult serum, sodium may be given as 140 millimols/1, whilst glucose may be given as 80 milligrams/100ml or an enzyme has the normal value of 5 units/ml. As a result, the clinician is expected to remember many different figures representing the normal values of body fluid chemicals and their associated normal ranges so that he can use laboratory analyses effectively for the treatment of his patient. In fact, when one examines how the clinician actually uses laboratory results, one often finds a much cruder use than is implied by the precision of the test figures. Usually the clinician is only interested in whether or not a value is outside the normal range and if so, whether it is slightly or distinctly outside that range.

Some years ago, therefore, I proposed a system of presenting chemical results in terms of their deviation from the normal population mean values using a scale of units derived from the associated normal ranges. This system accepts that the clinician is primarily interested in how far his patient's chemical 'parameters' have moved from normality, not so much in the absolute levels themselves. In place of absolute values, the mode of the population of normal results for any particular chemical is given the value of zero units, the lower end of the normal range −1.00 units, and the upper +1.00 units; the absolute value of the chemical determination is then re-expressed in this scale of 'range units'. As an example: the normal adult serum proteins have a mean concentration of 70g/1 and a normal range of ±10g/1; on the new scale a normal value of 70g/1 would be read as 0.00 units, a low value of 50g/1 as −2.00 units. The conversion, of course, is easy with the help of a small computer or hand calculator and account can be taken of deviation from normal gaussian distributions which occur with some blood chemicals. This new scale facilitates the reading of chemical results for the clinician since it allows him to recognise immediately whether or not a chemical level is within normal limits. As soon as a figure is found to be below −1.00 units or above + 1.00 units, it is recognised as being outside the normal limits, and the number of units will indicate the extent to which this particular chemical is departing from normality in its concentration. In place of a multitude of different normal values and ranges for different chemicals, we have an identical set of values for all.

This very simple conversion allows a new approach to the interpretation of the significance of multiple analyses, those which occur when we analyse a single sample for several different chemicals. The analysis of a particular blood chemical can be regarded as a point on a line, the latter being a one-dimensional scale where the normal population of such results cluster around the origin at 0.00 units and extends to ±1.00 units either side. The diabetic, for example, might have a blood glucose level which is distinctly high and this could be represented by a point on the right-hand end of the line shown in

Figure 2.1; a hypo-glycaemic level, on the other hand, would be represented by a point on the line to the left beyond the value of –1.00 units. If two different blood chemicals are determined in a single sample, the two values could be regarded as representing a point (p, q) in a two-dimensional space, one axis of which represents the values of one chemical parameter and the other axis representing the other (Figure 2.2). The position of this point in this space represents in some sense the condition of the patient, his deviation from normality as represented by the origin (0.0). If the line (0.0) to (p, q) is considered, its direction can be represented by its point of interception with the unit circle at (m, n). (In passing, it should be noted that, in such a space, points representing the normal population of pairs of chemical results do not form a perfect circle, nor do they form a perfect square as implied by 'one-dimensional' analyses. We have found that they tend to form 'super circles' of formula $R^n = X^n + Y^n$ where n = 2.5 to 4.0.) What we have observed is that, with chemical substances which do not transfer too rapidly from one compartment of the body to another, the direction of this line, as given by the co-ordinates of the point of intersection (m, n), can be correlated surprisingly well with the disease which is causing the patient to deviate from normal. Clearly, in such analyses there is a degree of float, but determinations on many patients with the same disease have, in general, shown the expected statistical clustering of the points; in other words, the general direction represents the diagnosis. The length of the line (0,0) to (p, q) represents how 'ill' the patient is at the time of the analysis; the longer it is, the more ill he is. This can be seen quite dramatically in some cases by watching how points move out along a fixed direction as the patient becomes more sick and how they move back in the opposite direction into the area of normality as recovery takes place. In cases of profound disease a point is sometimes reached in the outwards migration where the patient appears to have considerable difficulty in moving back into the normal range. Some physiological catastrophe appears to occur; the point wanders off its previously fixed direction and it seems as though it can no longer move back to normality – in other words, we have a dying patient on our hands. We have been investigating the limits at which these catastrophes occur; clearly the ability to predict the patient's approach to them could be of considerable clinical value.

This simple model can, of course, be expanded to include a large number of different chemical 'parameters'. We can obviously deal with three parameters by moving into a three-dimensional space; the point representing the three simultaneous analyses now has three co-ordinates

Figure 2.1 Adult blood glucose levels.

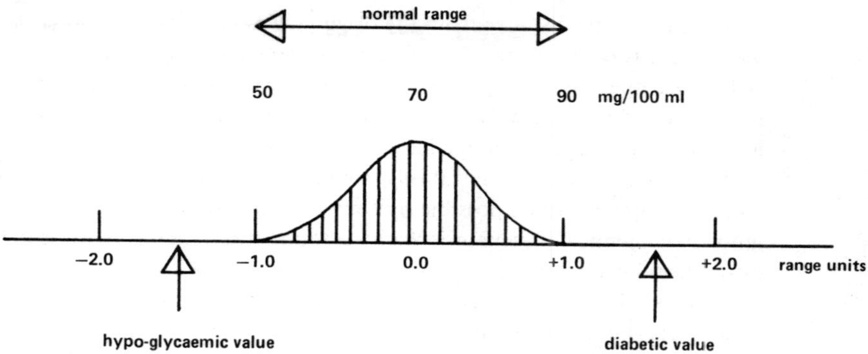

Figure 2.2

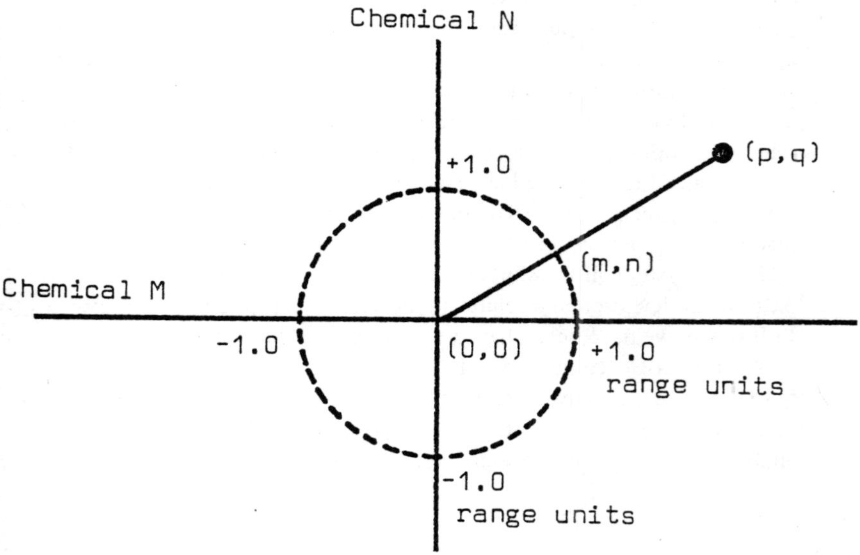

in place of two, and the direction of its deviation from normality is described, also by three co-ordinates in place of two, by the intersection of the line from the origin to the point through the unit sphere in place of a circle, and its distance from the origin, always a one-dimensional measure, is still a measure of the degree of deviation from normality. With these operational requirements it is very easy to move into a multi-dimensional space and we have examined up to 20 'parameters' in such studies. As a result, we have been able to form three conclusions.

Such a system detects deviations from normality at a much earlier stage than appears to be possible by simple examination of the individual chemical parameters. We have found, for example, significant differences between the patterns of the total plasma amino acids of normal and alcoholic subjects even though individual values of these amino acids all lay within normal limits. We can now recognise the distribution of different immuno-globulins in the serum of a child with a bad infection, even though these too may be individually within normal limits for a child of that age and environment. It appears, then, that the pattern or direction of deviation can be set up very early in the development of a disease; this has proved to be useful in helping a clinician choose between his 'hunches' at an early stage of disease.

By considering subsets of the parameters in a particular disease, it is comparatively easy to test which determinations appear to be the most useful for diagnostic purposes. This has indicated that the blunderbuss approach of the auto-analyser may not always be economically justified. It has also shown that some determinations regularly demanded by clinicians do not have a high correlation with the proposed diagnoses. When the poor discriminatory value of such tests has been pointed out to clinicians, the response has often been one of intellectual acknowledgement but with demands subsequently continuing to be made for the analysis 'just in case it might prove useful after all'!

Such a system allows simple sorting by computer of observed patterns of deviation and relates them with the proposed diagnoses. In this way we have been able to determine the diagnostic implications of a number of patterns of results which have not been noted before; in other words, we have been able to contribute to the more efficient teaching of medical students by showing that particular chemical analyses can help in determining particular diagnoses.

2.3 Mathematics in Scientific Research and Education

B.V. Gnedenko

The place of mathematics in contemporary scientific research has grown steadily more important. At the same time, the responsibility of the mathematician for the success of research as a whole has increased. One can consider the following as aspects of the mathematician's role in research into natural processes:

participating in the formulation of a mathematical model of the phenomena;
developing a mathematical theory on the basis of the model which has been formulated;
obtaining corollaries and both qualitative and quantitative deductions from the theory;
participating in the planning of a programme of experimental work for the verification of the theory, including the specification of those parameters of those processes which must be controlled;
modifying the model if discrepancies between experiment and theory appear to be significant.

The following example illustrates these aspects. A few years ago I was approached by astronomers who asked me to give them an answer, however approximate, to the following question: 'If we were to build an astronomical station on the surface of the moon would it be possible to calculate whether after 15 to 20 years of cosmic dust particle erosion the optical surface of the instruments would still have an optical transparency of at least 50 per cent?' In the above example the mathematical model had to anticipate a natural experiment. The researcher had data from observations of the flights of space stations and the results of a few laboratory investigations which permitted him to make an estimate of the intensity of the stream of cosmic dust, the distribution and dimensions of the particles and their speed. On the basis of the data at my disposal I constructed two simple models giving estimates of the probability of preserving optical transparency from above and below. Naturally both models are subject to further analysis. In this example computers are used only for calculations.

This outlines the place of mathematics in scientific research. In the light of this, what should be the mathematical education of the scientific experimenters is not an end in itself but a means of obtaining knowledge. In this connection the scientist's mathematical education must contain:

- a minimum of general mathematical knowledge;
- the presentation of basic mathematical models of phenomena in a given area of scientific knowledge;
- a theory of experimental design and analysis of experimental data;
- illustrations with very simple examples of how mathematical models are constructed and how they are evaluated in comparison with the experiment;
- training in the critical analysis of proposed mathematical theories of phenomena and other experimental verification.

At the second stage of training, there should be specialist seminars led jointly by representatives of a particular area of knowledge and by mathematicians. The aim of these seminars would be to demonstrate mathematical methods in action and to establish the habit of systematic collaborative work, in which each has his own share of the work. Naturally, not only analytical models, but also the modelling of processes on computers should be discussed in these seminars. Such seminars definitely require an element of discussion, in order to introduce young people to the practice of scientific investigation.

It is also important to consider the combination of education and research. There is no doubt that university teachers should be obliged to combine their teaching with active scientific research. The teacher not involved in scientific work will not know how to guide the young people whom he teaches and on the other hand the researcher who is not teaching will not be able to share the experiences he has gained from research with students so much in need of an experienced instructor.

At Moscow University there is a basic course which aims to attract students to do research and involves seminars at which the following activities take place;

- students present review papers;
- seminar leaders discuss each paper with the aim of clarifying those basic propositions fundamental to the proposed conclusions and drawing out new problems for consideration;
- students suggest answers to the questions posed by these problems and these are then considered by all the participants.

In the seminar I lead on queueing theory, engineers, economists and organisers of production take part regularly; less regularly medical students, biologists and physics students participate. We consider the systematic critical analysis of problems and proposed solutions presented by specialists of other fields of knowledge to be very important for students and for the other participants.

As a result of this approach a significant number of independent pieces of work are published every year by participants. These are joint publications by mathematicians and applied scientists.

2.4 Some Difficulties

It is, however, important to appreciate that, even given a basis of mathematical awareness on the part of biomedical scientists, the problem is not entirely solved by creating a team of mathematicians and scientists to work on a particular piece of research.

Despite a willingness to use quantitative methods, the scientist may not be able to see clearly what results are required or attainable. It is also possible that the most appropriate piece of mathematics for the task has yet to be formulated. The problem facing the mathematician is two-fold. First, he will have to understand the scientists' experimental model. Then, by abstraction and simplification, to explain to the scientist what limitations he has found necessary to introduce in his abstract view of the same experimental model.

The difficulty of description results from the way a mathematician will approach and analyse a given problem. Typically, he will abstract from all the technical details an overall picture of what seems essential. Then he will simplify his set of hypotheses or assumptions and proceed to prove that they are correct. However, rigorous proofs are seldom possible within the existing mathematical framework, unless some further mathematical and/or logical simplifications are introduced.

It is at this point that the mathematician often finds himself isolated and can get no sympathy from his science colleagues unless he is able to express himself in their technical language. What seems to be needed is a mechanism to introduce the abstract modelling concepts into a formal human-like language where both mathematicians and scientists can be trained alike and communicate with a greater efficiency and rigour.

3 COMPUTERS IN LIFE SCIENCE RESEARCH

3.1 A Review of Applications

Despite the difficulties referred to in the previous chapter, many examples exist of the important collaboration between computer scientists, mathematicians and life scientists in tackling multidisciplinary problems.

During the last few years computers have played an increasingly important role in science and clinical medicine; in the future their importance as a scientific tool is likely to increase still further. At the present time, computers are used widely in the biological sciences in fundamental research work, while in clinical medicine their current applications are largely in the area of patient record keeping and simple data logging. This is certain to change within the next few years because techniques which are presently being employed for data analysis and modelling of biological systems will be increasingly employed in the clinic for diagnostic purposes.

At present the most important clinical applications of mathematics and computers lie in the cardiovascular field. Two areas have gained prominence; ECG analysis, and more recently, computer modelling of the cardiovascular system. The prompt recognition of dysrhythmias in the coronary care unit by electrocardiographic monitoring has produced a very significant reduction in hospital mortality. As a result, a number of methods for automatic surveillance have been developed which rely on computers. Another area of increasing interest is surface mapping; isopotential maps of patients' thorax are computed in order to detect cardiac abnormalities. The growing interest in surface mapping would seem to derive from the clinicians need to be able to visualise the clinical problem. This is an important factor in the application of computers in clinical medicine because while physical scientists and engineers are happy to consider problems in terms of hard data, clinicians are trained to work in terms of soft data and visual representations. Consequently, a major part of computer processing of data in clinical medicine in the future is likely to be the conversion of hard data obtained from the automatic analysis procedures into soft data for presentation to the clinician.

Another important area with tremendous potential is computer modelling. Models have been used up to the present time in the study of various physiological systems, although until now this work has been exclusively within pure research. Models do however, provide a means of assessing clinical parameters. The normal approach is for the computer to receive input signals from the physiological system under investigation, and these signals are simultaneously compared with similar signals obtained from a computer model of the system. The differences between corresponding variables are calculated and the model parameters adjusted automatically to minimise the overall difference. After a number of iterations the difference becomes small and the computer model is then said to be an image of the physiological system. Because of the various limitations which are placed upon diagnostic procedures, many important pieces of data are often unobtainable. However, when a computer image of a patient's physiological systems has been created, the image can be interrrogated in detail; hence, the method has enormous clinical potential.

Techniques of this type are already being used in the United States in the treatment of respiratory failure. In these hospitals, treatment, in the absence of physical damage, is seen in terms of moving the patient's respiratory system from the abnormal state in which it resides, back to its normal state. Such a transition is almost certainly multivariate in nature, and consequently it is essential that this fact be considered when treatment strategies are being developed. The use of multivariate analysis requires the use of fairly sophisticated mathematical and computing techniques; thus this application represents the increasingly important collaboration between mathematics and the physical sciences, and the life sciences. It is to be hoped that in the future such collaboration will provide the clinician with new and far more powerful diagnostic tools.

3.2 Mini Computers Interfaced to Gamma Scintillation Cameras in Nuclear Medicine: The Role and Achievements in the Development of Nuclear Cardiology

V.B. Bosnjaković

Nuclear medicine represents one of the 'prototype' multi-disciplinary scientific fields and mostly owing to this fact, it is one of the fastest expanding areas today. Advances made in radiochemistry, the introduction to medical practice of short-lived radio-tracers and a number of radiopharmaceuticals (technetium ^{99m}Tc and others), and in particular the development of specific electronic instrumentation which enabled efficient utilisation of these agents, have caused fast growth of this new medical discipline.

One of the major steps forward in modern nuclear medicine was made in the field of instrumentation when H. Anger invented the gamma scintillation camera (GSC).[1] The Anger model, despite the subsequent introduction of others,[2] is at present the most widely used type of GSC; it is in essence a complex and sophisticated scintillation detector, an ingenious innovation based on achievements in physics, electronics and computer science. Gamma scintillation cameras as 'fast' imaging devices had a strong impact on clinical research and diagnostic *in vivo* nuclear medicine procedures. They provided an instant visualisation of the distribution of radio-tracers in specific organs and tissues, enabling not only efficient static morphological visualisation but in particular dynamic imaging leading to development of a wide range of qualitative and quantitative diagnostic and research applications.

The essential performance parameters of Anger cameras – intrinsic spatial resolution, energy resolution and temporal resolution – have continuously improved during the last 10 to 15 years. At present the intrinsic resolution in space is about 3 mm, energy resolution less than 15 per cent and temporal resolution around 1.5-2 microseconds;[3] such a short 'dead' time allows counting rates as high as 200,000 counts/sec without loss. These capabilities – which made it possible to perform fine morphological and fast dynamic studies – 'naturally' required interfacing of mini-digital computers to GSCs for adequate processing, analysis and display of the output signals.[4,5,6] Characteristic positioning (X and Y) signals of the distribution of radioactive events in the specific organ (tissue) and energy coding (Z) signal, as analog voltages from Anger cameras, are directly amenable for on-line analog-to-digital conversion and processing by a digital computer. Actually, in modern nuclear medicine, a GSC interfaced to mini-digital computer system is considered the whole instrument (Figure 3.1).

Such a complex machine revolutionised particularly the fast dynamic studies of the central circulation, enabling introduction and development of the new non-invasive technique – radionuclide cardioangiography – which provided the broadest spectrum of qualitative and quantitative information on the central circulation.[7] Owing chiefly to these innovations in dynamic *in vivo* studies, nuclear cardiology as a new field in nuclear medicine (incorporating also *in vivo* peripheral circulation studies and relevant *in vitro* radioimmunoassay

Figure 3.1 Schematic presentation of an Anger gamma scintillation camera as interfaced to the dedicated mini-digital computer system, including analog-to-digital converters (ADC), computer-central processing unit (CPU), display, control terminal and printer, disc and tape storage.

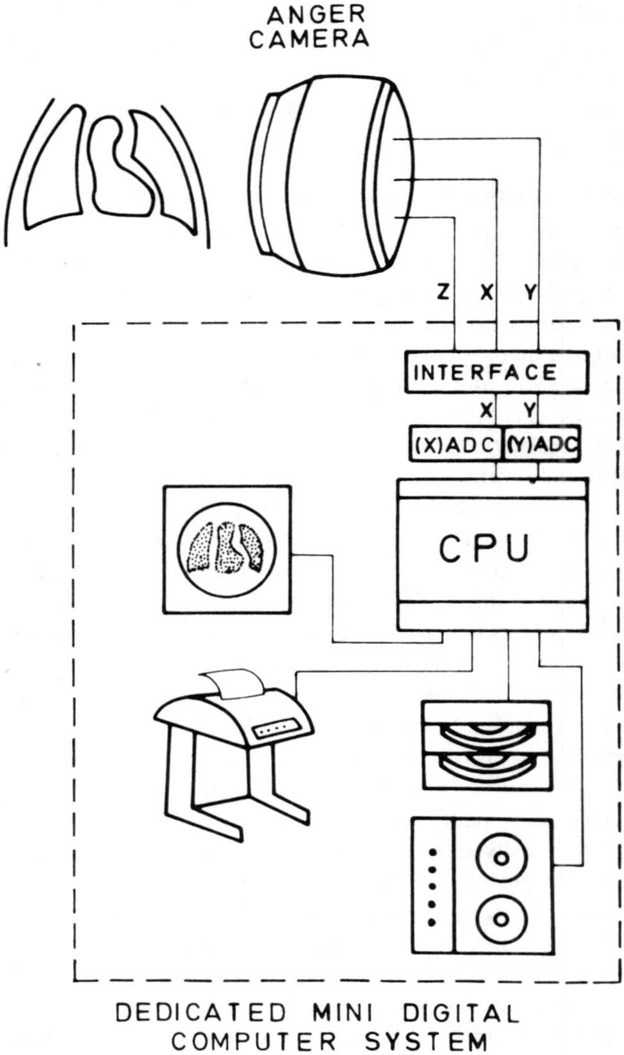

and other studies) has grown rapidly, being in the last three to five years accepted in modern diagnosis by clinical cardiologists. At present, non-invasive *in vivo* nuclear cardiology procedures are the most representative and attractive examples of the benefit of the use of the mini-digital computer – GSC systems in nuclear medicine. The following procedures have gained a reputation of particular merit in clinical diagnostic cardiology: (1) qualitative and quantitative radionuclide cardioangiography; (2) detection of intracardiac shunts; (3) ECG gated cardiac blood pool imaging; (4) acute infarct imaging; (5) myocardial imaging.

Qualitative and Quantitative Radionuclide Cardioangiography

This in essence is a technique in which the first transit of an intravenously injected bolus of radioactive tracer through the heart chambers is detected with a GSC placed over the chest wall. Being similar to X-ray contract cardioangiography, it was in its early stage (without the aid of a digital computer) a qualitative technique limited to dynamic imaging. With the specific attachments to GSCs to obtain fast sequential imaging (35 mm film cameras, variable time-lapse videoscintiscope,[8] magnetic videotape recorders[9]), such a dynamic visualisation provided satisfactory spatial and temporal resolution for qualitative visual analysis and diagnosis of acquired valvular heart diseases, some congenital malformations, ventricular and aortic aneurisms.[10] The logical improvement to utilise completely the 'direct digital' specificity of a suitable radioactive tracer and of consequent output GSC analog signals was the attachment of mini-digital computers to modern GSCs in order to make the technique a quantitative one. The quantitative approach is based on the computer generating time-activity curves from assigned 'regions-of-interest' (ROI) over specific cardiac chambers during playback analysis of the stored data. Interpretation and analysis of these time-activity histograms for obtaining the quantitative parameters by adequate computer programming, theoretically stems from the Stewart-Hamilton dilution principle and specific quantitative relations combined with regional transit-time analysis. Thus, it has been shown feasible to obtain in a non-invasive way a wide spectrum of quantitative haemodynamic parameters: cardiac output, forward stroke volume, mean transit time of the right heart, end diastolic volume and ejection fraction of the right ventricle, mean pulmonary transit time, pulmonary blood volume, ejection fraction of the left ventricle, end diastolic (and end systolic) volume of the left ventricle, and regurgitation fraction in valvular malformations. The formulas of haemodynamic parameters obtainable by quantitative radionuclide cardioangiography are listed in Figure 3.2.

Figure 3.2 Formulas of haemodynamic parameters obtainable by quantitative radionuclide cardioangiography; CO — cardiac output; Ceq — equilibrium concentration on first transit-time-activity histogram; BV — blood volume; FSV — forward stroke volume; p' — pulse rate; $\overline{MRHTT}$ — mean right heart transit time; EDV — end diastolic volume; EF — ejection fraction; $\overline{MPTT}$ — mean pulmonary transit time; t max LA — time of occurrence of the peak of left atrial time-activity histogram; PBV — pulmonary blood volume; ED — end diastole and ES — end systole (corresponding values on the ventricular high frequency first transit-time-activity histogram); RH = right heart; RV = right ventricle; LV = left ventricle; (AV) = average.

QUANTITATIVE RADIONUCLIDE CARDIOANGIOGRAPHY

$$CO = \frac{C_{eq}}{\int_0^\infty C(t)dt} \cdot BV$$

$$FSV = \frac{CO}{P}$$

$$\overline{MRHTT} = \frac{\int_0^\infty t\, C(t)dt}{\int_0^\infty C(t)dt}$$

$$EDV_{RH} = \overline{MRHTT} \cdot CO$$

$$EF_{RV} = \frac{FSV}{EDV_{RH}}$$

$$\overline{MPTT} = t_{max.} LA - \overline{MRHTT}$$

$$PBV = \overline{MPTT} \cdot CO$$

$$EF_{LV} = \frac{ED_{LV} - ES_{LV}}{ED_{LV}}$$

$$EDV_{LV} = \frac{FSV}{EF_{LV}}$$

$$EDV_{LV(AV)} = \frac{\pi}{6} LM^2$$

$$FR = \frac{(EDV_{LV(AV)} \cdot EF_{LV}) - FSV}{EDV_{LV(AV)} \cdot EF_{LV}} = \frac{EDV_{LV(AV)} - \frac{FSV}{EF_{LV}}}{EDV_{LV(AV)}} = \frac{EDV_{LV(AV)} - EDV_{LV}}{EDV_{LV(AV)}}$$

Detection of Intracardiac Shunts

This has been one of the most attractive problems in nuclear cardiology since the period before the introduction of GSCs. Modern nuclear cardiology diagnosis of intracardiac shunts, despite being based on radionuclide cardioangiography, still has some specificities. A qualitative visual approach, as done in the way mentioned above, was found unsatisfactory for the diagnosis of more frequent left-to-right (L-R) shunts.[11] Therefore, the other approach based on the analysis of time-activity curves (derived from ROIs) has been developed; it enabled quantification of shunts as well, by determining the specific parameter, pulmonary-to-systematic flow ratio (Qp/Qs). Two different principles based on time-activity histograms exist. One is based on the first transit-time-activity curve derived from a lung field and is limited to detecting L-R shunts only.[12]

The other one, a comprehensive technique for the detection of both right-to-left (R-L) and L-R shunts, will be dealt with in more detail.[13-16] It assumes obtaining simultaneous first transit-time-activity curves from both atria or/and both ventricles. In such a situation single peaked curves obtained over the right and relevant left heart chamber, with peaks appearing separately at corresponding time intervals, indicate a normal situation without shunting. The pattern in an R-L shunt case is a single peaked curve over the right heart and characteristic double-peaked curve over the left heart, the first peak of which, corresponding to the time interval of appearance of the peak over the right heart, is due to an R-L shunt. In a situation with L-R shunt, the characteristic pattern is a double-peaked curve over the right heart chamber the second peak of which, occurring simultaneously with the normal single peak over the left heart, is due to the L-R shunt. Thus, time inter-referred peaks of the simultaneous time-activity curves assure the qualitative diagnosis of shunting. By this approach it is also possible to discriminate the occurrence of shunting on either the atrial or ventricular level; if, in the case of an L-R shunt, the described pattern is found on both atrial and ventricular level, this would indicate the existence of an atrial septal defect (ASD), whilst the appearance of an L-R shunt pattern only at the level of ventricles is indicative of an isolated ventricular septal defect (VSD). A particular advantage of this method is that the characteristic double-peaked curves are directly amenable to quantitative analysis for providing the Qp/Qs ratios; if by the extrapolation of downslopes (using exponential function fitting) and 'peeling off' techniques the areas under specific peaks are determined, then the relations which provide the Qp/Qs parameter in R-L and L-R shunts exist as shown in Figure 3.3.

Figure 3.3 shows the schematic presentations of haemodynamic situations in R-L and L-R shunts, with the monitoring sites, ROIs and the characteristic double-peaked curves generated in R-L shunt from a left heart chamber (LHC) and in L-R shunt from a right heart chamber (RHC). In the case of an R-L shunt proportionality is demonstrated between the area under the first peak (A_{RLS}) to the R-L shunt flow and between the area under second peak (A_L) to the pulmonary flow, resulting in the relation for Qp/Qs. In the case of an L-R shunt there is evidence of proportionality between the area under the first peak (A_R) to the pulmonary flow and of the area under the second peak (A_{LRS}) to the L-R shunt flow, giving the corresponding relation for Qp/Qs. A particular theoretical advantage should be pointed out; the geometrical factor which is critical in haemodynamic measurements with scintillation probes vanishes in this analysis, since the inter-related areas of a double-peaked curve are obtained from the same ROI and are 'seen' with identical geometry.

Such a comprehensive method was a challenge to develop an automated diagnosis of shunts by a (dedicated) digital computer.[17] The computer program has been written to provide at first qualitative analysis of shunting based on the synchronous (or not) time sequences of appearance of characteristic peaks. When this has been established, the areas under the peaks, as analytically determined, are calculated and an algorithm finally provides computation of Qp/Qs for either the R-L, L-R or bidirectional shunting, based on the formulas given in Figure 3.3. Thus, the final diagnosis as determined by the computer provides the qualitative diagnosis of whether the shunt is R-L, L-R or bidirectional, the quantitative parameter Qp/Qs ratio and in case of an L-R shunt the clinical diagnosis of ASD or VSD, the last being based on the separate entrance information on whether the shunting occurs on both atrial and ventricular level, or only at the ventricular level. The example of such an automated non-invasive diagnosis is given in Figure 3.4 in which the computer has found that the shunt is L-R, clinical diagnosis VSD and Qp/Qs = 1.76/1. Cardiac catheterisation in this case provided the same diagnosis with Qp/Qs being 1.9/1.

Figure 3.3 Scheme of haemodynamic situations which exist in R-L and L-R shunts (Refs. 15, 16); Qp – pulmonary flow; Qs – systemic flow; Qrls – R-L shunt flow; Qlrs – L-R shunt flow; Al – area under 'normal' left heart chamber peak; Arls – area under R-L shunt peak; Ar – area under 'normal' right heart chamber peak; Alrs – area under L-R shunt peak.

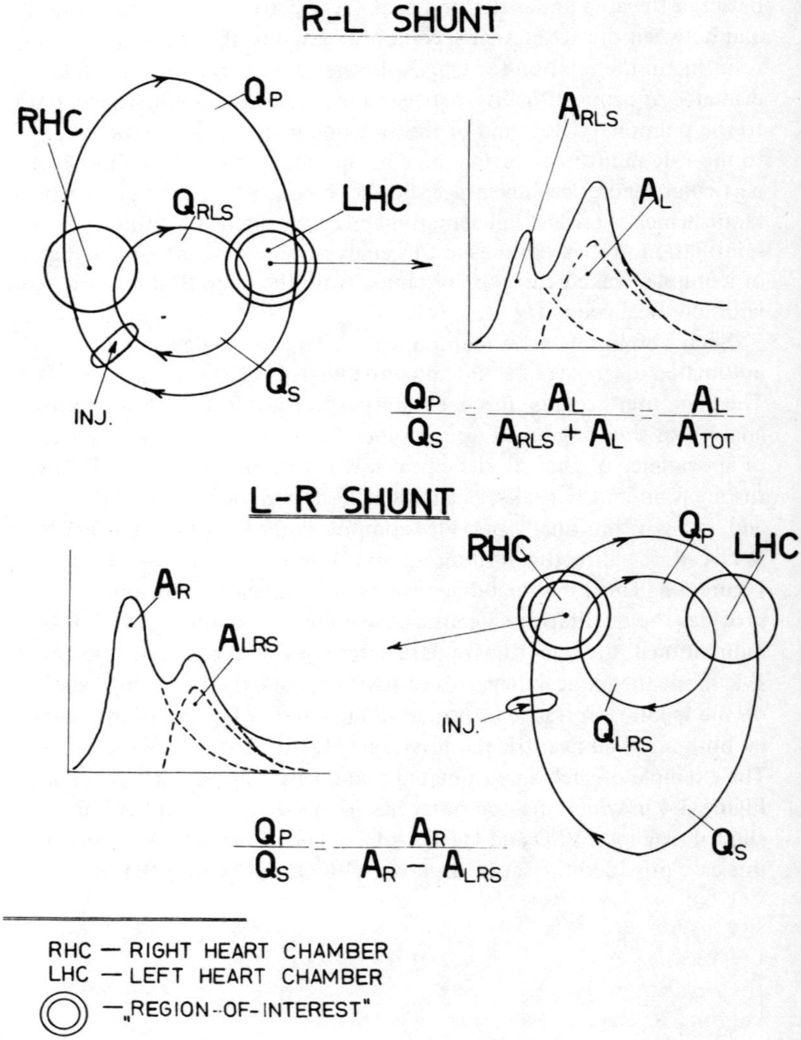

Figure 3.4 Automated non-invasive diagnosis of intracardiac shunts (an L-R shunt case) as inscribed by the adequately programmed digital computer (Ref. 17); VSD – ventricular septal defect; Qp/Qs – pulmonary/systemic flow ratio.

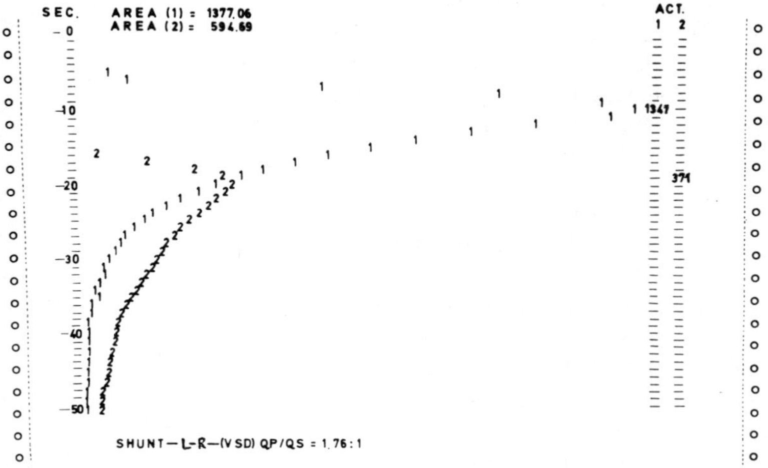

ECG Gated Cardiac Blood Pool Imaging

This is a nuclear cardiology technique which has gained relatively wide acceptance in diagnostic cardiology in the 1970s.[18] It uses a radioactive tracer which does not clear rapidly from the circulation after its first transit through the heart cavities. Tracers, like technetium^{99m}Tc, label human serum albumin or red blood cells. This is possible because the carriers are homogeneously mixed within 5-10 min. of intravenous infusion enabling the imaging of cardiac blood pools or cavities. Images of the right and left ventricle in end-systole (ES) and end-diastole (ED) are produced, as the synchronous ECG triggers a GSC to accumulate counts and form the images only during downslope of the T-wave (end-systole) and P-Q interval (end-diastole). Approximately 600-800 cardiac cycles are accumulated and recorded during 7-10 minutes. Since the tracer remains in the circulation, images can be done in any suitable projection, usually in left anterior oblique and right anterior oblique, and stored by a digital computer. Computation for obtaining the magnitudes of the left ventricular and end-systolic and end-diastolic volumes is based on the area-length principle known from biplane

X-ray contrast angiography. Such an approach (formulae listed in Figure 3.2) provides the value of an 'average' end-diastolic (or end-systolic) volume, which is used for the calculation of ejection fraction; or, in cases with valvular malformations, by inter-relating 'average' end-diastolic volume (EDV) to the EDV obtained on the basis of forward stroke volume, for computation of regurgitation fraction (formulae in Figure 3.2). The simple visual inspection of dynamic playback or the appropriate computer processing of cardiac blood pool imaging provides the way of evaluation of the regional ventricular wall motion.

Acute Infarct Imaging

It has been shown that technetium ^{99m}Tc labelled pyrophosphate, introduced primarily as a bone imaging agent, concentrates in regions of acute infarction, thus being able to detect and localise it.[19] There is evidence that pyrophosphate (or polyphosphate) accumulates in predominantly irreversibly damaged myocardial tissue. Experience from the use of this popular agent in cardiological practice indicates that there is a probability of more than 90 per cent of visualising transmural infarction when the procedure is done from 24 hours up to 6-7 days after its occurrence. Imaging as done with an adequately positioned GSC over the chest wall is usually performed 1.5-2 hours after the injection of the radio-tracer. The size of anterior and lateral wall infarctions can be estimated by this method by the simple planimetry measurements of the area of imaged myocardial ^{99m}Tc pyrophosphate uptake. This technique became a standard method for confirming acute infarction with dependability similar to routine ECG and enzyme elevation findings.

A computer-generated multiradionuclide colour-coded imaging of myocardial infarction has been produced. If the first transit radionuclide cardioangiogram is displayed as a green image and the following pyrophosphate imaging as a red one; a third colour, yellow, can be computer generated in areas of superimposed activity. Thus, sizing and localisation of infarctions is greatly facilitated.[20]

Myocardial Imaging

This technique assumes the use of radio-tracers which localise in normal myocardial cells, visualising functional myocardial cell mass. Differing from the previous technique which concerns the use of tracers taken up by the damaged myocardial cells and visualises only acute infarctions as 'hot' spots, this technique produces 'cold' spots with intensities dependent on the degree of functional and morphological disturbances. 'Cold' (non-functional) or hypofunctional spots are therefore seen in areas ranging from the acute and chronic infarctions (fibrous tissue – scars) to the regional ischemias as being permanent or transiently induced by stress-exercise. Several radio-tracers have been used for this type of technique encompassing radioactive potassium ^{43}K and its physiologic analogs like radioactive isotopes of cesium (^{129}Cs), rubidium (^{84}Rb, ^{86}Rb, rubidium studies being done with scintillation probes, not as GSC imaging), thallium (^{199}Tl, ^{201}Tl).[21,22]

The most recently commercially available amongst those listed, viz. thallium ^{201}Tl, owing to its physical characteristics (photon energies suitable for studies with Anger cameras; physical half-life of 73 hours allowing for convenient supply) and physiological properties (evidence that is retained in myocardium longer than potassium) became the most popular and is the only one used at present for this type of study.[23] Besides the detection of infarctions and scar regions, a particular application should be pointed out: the use of ^{201}Tl in the non-invasive functional stress test[24,25] (as previously done with ^{43}K). Thallium imaging is performed at rest first and then a week later (allowing for disappearance of remaining activity) the repeated imaging at the peak of (tread-mill or bicycle ergometer) exercise. This reveals various stages of coronary insufficiency, and a 'cold' spot appears in an underperfused region or increases its size compared to the test at rest. The recent experience from a number of reports evaluating ^{201}Tl exercise myocardial imaging has evidenced that the thallium stress test exceeds the sensitivity and specificity of an ECG stress test.

The computer processing and retrieval of thallium imaging has shown advantages in evaluating the size of functional myocardial mass. One of the most recent developments is the technique of using dual isotope colour-coded computer-produced motion pictures.[26] Thallium ^{201}Tl myocardial wall and technetium ^{99m}Tc albumin heart blood pool combined GSC images, using ECG synchronisation, are computer acquired. Two isotopes are shown in contrasting colour, regions of wall and pool superimposition in a secondary colour and all displayed in real-time cine mode. The technique is superior to static images for the detection of myocardial perfusion abnormalities since it delineates the extent of infarctions more precisely than thallium imaging alone, providing correlation between perfusion defects and corresponding pool edge dyskinesis. It has also been shown that the exercise-induced defects are best seen in diastole (Figure 3.5).

Figure 3.5 Schematic black and white presentation of the technique using dual isotope colour-coded computer-generated motion pictures for assessing the cyclic changes in myocardial defects; 201-Thallium (TH) is given in green, 99m-Tc albumin blood pool (BP) in red, overlap in yellow; left anterior oblique position of the left ventricular area in systole and diastole shows an apical defect best seen in diastole (Ref. 26).

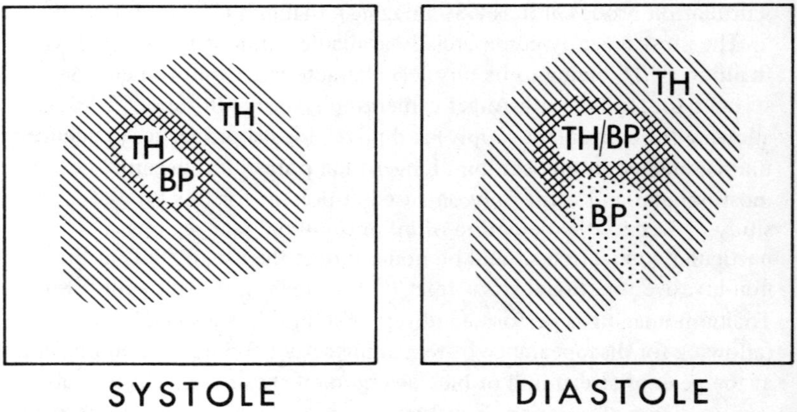

Concluding Comments

The instrument characteristics for nuclear cardiology dynamic studies are critically important since haemodynamics of the central circulation constitute the fastest flow studies, where the available counting time is about 15 to 30 seconds. For precise analysis of individual cardiac cycles, counting intervals have to be no longer than 0.1 seconds. For providing optimum counting statistics in such short intervals, the high counting rates obtained from administering adequate amounts of radioactivity (currently available from short-lived radionuclides) are required. This emphasises the necessity of not only the optimum temporal but also spatial resolution. Thus, a modern 'fast' and spatially high resolutive GSC, as pointed out above, is the basic prerequisite. It has been made mobile to approach patients in coronary care and in intensive care units and to perform at the bedside the non-invasive diagnostic procedures. The overall weight of the mobile GSC has been reduced owing to the fact that the most frequently used radio-tracers are the low-energy gamma-emitting isotopes (^{99m}Tc, ^{133}Xe, ^{201}Tl, ^{123}I) which do not require heavy lead shielding and collimation.

The large amount of data produced during the GSC cardiac studies can be practically utilised only with a recording system which permits retrieval and manipulation of data. This has been provided by interfacing mini-digital computers to gamma scintillation cameras. 'Non-mobile' GSCs located in nuclear medicine departments are directly connected (interfaced) to 'central' mini-digital computers. Mobile GSCs are equipped with compatible magnetic tape or disc storage devices for later playback and processing of data by these 'central' (still dedicated) machines. The basic characteristics and capabilities of mini-computers which are considered satisfactory in contemporary nuclear medicine are defined, as would be expected, according to nuclear cardiology requirements. 'Fast' analog-to-digital converters (ADC) of 200 MHz, 64 × 64 and 128 × 128 matrices, a 16 bit word with 32-48K words memory and 7 million to 14 million bytes disc capabilities are required.

Two quotations characterise the 'state of the art' and future prospects of the topics reviewed in this article.

> A strong interaction exists between the adjectives 'quantitative' and 'digital'. One of the important developments of instrumentation which enables Nuclear Medicine to become quantitative is the development of the digital computer to the stage where it is fast enough, small enough, cheap enough, and responsive enough for ordinary mortals to deal with on a friendly basis. . .Those in the field of computer science have little doubt of the capacity of the computer to revolutionise many fields, including nuclear cardiology.[27]

> The successful developments in radionuclide cardioangiography within a few short years provide an insight to our future. The ability to visually evaluate ventricular wall motion, to measure ventricular ejection fraction, to. . .all by means of a safe, non-invasive technique, exemplifies what can be accomplished by the co-ordinated efforts of basic and clinical scientists.[28]

References

1. H.O. Anger, 'Scintillation Camera', *Revue of Scientific Instruments*, 29, 27, 1958.
2. M.A. Bender and M. Blau, 'The Autofluoroscope', *Nucleonics*, 21, 52, 1963.
3. G.J. Hine and P. Paras, 'Performance of Scintillation Cameras', *Journal of Nuclear Medicine*, 16, 1206, 1975.
4. H.N. Wagner Jr. and T.K. Natarajan, 'The Computer in Nuclear Medicine',

CONF 710425, Proceedings Symposium on Sharing of Computer Programs and Technology in Nuclear Medicine (Oak Ridge, Tenn., 1971).

5. P.M. Weber, L.V. dos Remedios and I.A. Jasko, 'Quantitative Radioisotopic Angiocardiography', *Journal of Nuclear Medicine*, 13, 815, 1972.

6. R.H. Jones, L. Lidofsky, T. Budinger and P. Weber, 'Instrumentation', in *Quantitative Nuclear Cardiography*, R.N. Pierson, J.P. Kriss, R.H. Jones and W.J. MacIntyre (eds) (Wiley, New York, 1975), p.231.

7. H. Parker, P.M. Weber, D.C. Van Dyke, H. Davies, P. Steele and R. Sullivan, 'Evaluation of Central Circulatory Dynamics with the Radionuclide Angiocardiogram', in *Cardiovascular Nuclear Medicine*, ed. H.W. Strauss, B. Pitt and H. Everette James Jr. (Mosby, St Louis, 1974), p.67.

8. J.P. Kriss, W.A. Bonner and E.C. Levinthal, 'Variable Time-lapse Videoscope: A Modification of the Scintillation Camera Designed for Rapid Flow Studies', *Journal of Nuclear Medicine*, 10, 249, 1969.

9. D.T. Mason, W.L. Ashburn, S.C. Harbert, L.S. Cohen and E. Braunwald, 'Rapid Sequential Visualization of the Heart and Great Vessels in Man Using the Wide-Field Anger Scintillation Camera; Radioisotope-Angiography Following the Injection of Technetium-99m', *Circulation*, 39, 19, 1969.

10. J.P. Kriss, L.P. Enright, W.G. Hayden, L. Wexler and N.E. Shumway, 'Radioisotopic Angiocardiography: Wide Scope of Applicability in Diagnosis and Evaluation of Therapy in Diseases of the Heart and Great Vessels', *Circulation*, 43, 792, 1971.

11. P.J. Hurley, H.W. Strauss and H.N. Wagner Jr., 'Radionuclide Angiocardiography in Cyanotic Congenital Heart Disease', *Hopkins Medical Journal*, 127, 46, 1970.

12. D.L. Maltz and S. Treves, 'Quantitative Radionuclide Angiocardiography; Determination of Qp:Qs in Children', *Circulation*, 47, 1049, 1973.

13. V.B. Bosnjakovic, L.R. Bennett, W. Vincent and C. Newton, 'A Method for Diagnosis of Intracardiac Shunts Based on Radioisotope Angiography', *Radioaktive Isotope in Klinik und Forschung*, 10 Band, Urban und Schwarzenberg, p.290, 1973.

14. V.B. Bosnjakovic, L.R. Bennett, L.D. Greenfield and W.R. Vincent, 'Dual-isotope Method for Diagnosis of Intracardiac Shunts', *Journal of Nuclear Medicine*, 14, 514, 1973.

15. V. Bosnjakovic, L. Bennett, W. Vincent and L. Greenfield, 'Quantitation of Intracardiac Shunts by Radioisotope Angiography', *Proceedings of First World Congress of Nuclear Medicine*, Tokyo, p.230, 1974.

16. V. Bosnjakovic, 'Non-invasive Qualitative and Quantitative Diagnosis of Intra-Cardiac Shunts', *Medicinska Istrazivanja*, Supplement, 2, 33, 1975.

17. V. Bosnjakovic, 'Non-invasive Method for Diagnosis of Cardiac Shunts with Processing by the Digital Computer', *Proceedings of the Second Yugoslav Congress of Congenital Malformations* (Belgrade, 1976), p.457.

18. H.W. Strauss, B.E. Zaret, P.J. Hurley, T.K. Natarajan and B. Pitt, 'A Scintiphotographic Method for Measuring Left Ventricular Ejection Fraction in Man without Cardiac Catheterization', *American Journal of Cardiology*, 28, 575, 1971.

19. R.W. Parkey, F.S. Bonte, S.L. Meyer, J.M. Atkins, G.L. Curry, E.M. Stokeley and J.T. Willerson, 'A New Method of Radionuclide Imaging of Acute Myocardial Infarction in Humans', *Circulation*, 50, 540, 1974.

20. I. Mena, S.H. Joe, J.M. Uszler and J.A. Jengo, 'Multiradionuclide Color-coded Myocardial Infarction Imaging: Diagnosis, Sizing and Localization', *Proceedings of the Sixth Symposium on Sharing of Computer Programs and Technology in Nuclear Medicine* (Atlanta, 1976), p.177.

21. E. Lebowitz, M.W. Greene, R. Fairchild, P.R. Bradley-Moore, H.L. Atkins,

A.N. Ansari, P. Richards and E. Belgrave, 'Thallium-201 for Medical Use I', *Journal of Nuclear Medicine*, 16, 151, 1975.

22. P.R. Bradley-Moore, E. Lebowitz, M.W. Greene, H.L. Atkins and A.N. Ansari, 'Thallium-201 for Medical Use II: Biologic Behaviour', *Journal of Nuclear Medicine*, 16, 156, 1975.

23. H.R. Schelbert, H. Henning, P. Rigo, D.M. Chancey, S.E. Halpern, R.A. O'Rourke and W.L. Ashburn, 'Considerations of Thallium-201 as a Myocardial Imaging Agent in Man', *Investigative Radiology*, 11, 163, 1976.

24. I.K. Bailey, L.S. Griffith, J. Rouleau, H.W. Strauss and B. Pitt, 'Thallium-201 Myocardial Perfusion Imaging at Rest and During Exercise', *Circulation*, 55, 79, 1977.

25. J.L. Ritchie, G.B. Trobaugh, G.W. Hamilton, K.L. Gould, K.A. Narahara, J.A. Murray and D.L. Williams, 'Myocardial Imaging with Thallium-201 at Rest and During Exercise', *Circulation*, 56, 66, 1977.

26. H. Bishop, 'A New Technique for Assessing Defects and Cyclic Changes in the Myocardium: Dual Isotope Color Coded Computer-Produced Motion Pictures', *Softwhere; A Publication of Medical Data Systems Corporation*, 4, 5, 2, 1977.

27. R.N. Pierson, 'Introduction', in *Quantitative Nuclear Cardiography*, ed. R.N. Pierson, J.P. Kriss, R.H. Jones and W.J. MacIntyre (Wiley, New York, 1975), p.1.

28. F.H. Deland, 'Editorial', *Journal of Nuclear Medicine*, 19, 1, 1978.

3.3 The Analysis of 'Thermal' Oscillations in Heart Rate Variability

R.I. Kitney

In the late thirties and early forties a number of workers[1] identified fluctuations in digit blood flow which they associated with thermal activity in the approximate range of fifteen to thirty seconds periodicity. These oscillations formed the basis of our analysis of the thermal vasomotor control system. The question which was asked was: what kind of control system could produce this type of oscillation? After a careful study of the data, the conclusion which was reached was that the control system must be of the non-linear oscillatory type, i.e. the type of bounded oscillations which were observed must emanate from a non-linear rather than a linear control system.[2]

In 1962 Minorsky discussed[3] the theory of non-linear oscillatory systems in considerable detail. At this stage in the research it was important to test our deductions relating to the nature of the control system and to this end the property of such systems, which Minorsky calls entrainment, was extremely important. Minorsky's theory of entrainment formed the basis of a set of experiments.[4] The purpose of these was to test the non-linear nature of the thermal vasomotor control system by attempting to entrain the thermal oscillations in digit blood flow with a periodic thermal stimulus. After a careful study of the physiological literature it was decided that the stimulus should be achieved by placing one hand alternately in two tanks containing water at different temperatures. (In the first set of experiments the temperatures were 18°C and 46°C.)

Consider the case of an experiment in which a 20-second stimulus was applied. In the experiment the subject was asked to place one hand in the cold tank for 10 seconds followed by 10 seconds in the warm tank, followed by 10 seconds in the cold tank, and so on. Variations in finger blood volume were recorded from a photoplethysmograph placed on the small finger of the opposite hand, and the ECG (single chest lead) was recorded on an FM instrumentation tape-recorder for the duration of the experiment (15 minutes).

In general, for different stimulus frequencies the digit blood flow in the opposite hand had a dominant frequency component at the stimulus frequency, indicating entrainment. The range of entrainment was found[4] to be about 8 to 80 seconds.

One area of interest as far as the data analysis is concerned lies in the presentation of the experimental data. In order to study the entrainment phenomenon, it is necessary to determine whether or not there is any evidence of the periodic thermal stimulus in the digit blood flow waveform recorded from the photoplethysmograph. Figure 3.6 shows the type of experimental results which are obtained. Figure 3.6a is the spontaneous digit blood flow waveform recorded from a healthy subject; Figure 3.6b is for the same subject but with a 20-second period thermal stimulus applied to the opposite hand. Each record shows blood flow recorded over 500 seconds. The problem of identifying entrainment is immediately obvious when the two plots are compared, although there are differences between the plots, it is extremely difficult to define exactly what they are. Consequently, other forms of data presentation are often more useful. This is the case for the experimental data shown here. The alternative approach which has been adopted in our work[5] is to calculate the power spectra of the blood flow data. Figure 3.7 illustrates the power spectra of the data shown in Figures 3.6a and 3.6b respectively. A comparison of these plots immediately reveals that there is a marked difference between them. Figure 3.7a comprises a number of frequency components spread throughout the range 0-0.1 Hz, while Figure 3.7b has a single large component at 0.05 Hz (20-second period) which corresponds to the thermal stimulus frequency. Figure 3.7b provides far clearer evidence of entrainment than is possible with the data plotted in the conventional manner.

Further experiments led to a detailed analysis of the thermal vasomotor control system. The next stage in the research was to study the effect of these thermal stimuli on heart rate.

Two earlier papers were published on this aspect of the work; one on the nature of spontaneous rhythms in physiological control systems,[6] and the other[7] on the analysis of heart rate variability specifically. The essence of these papers as far as heart rate variability is concerned can be summed up in two statements:

1. There appeared to be three frequency components in the normal heart-rate variability power spectrum which arise from the activity of physiological control systems: a temperature component in the range 0.05 Hz, a blood pressure component at around 0.1 Hz and a respiratory component at about 0.25 Hz, depending on the respiration rate. (These components can be observed during spontaneous resting conditions.)
2. These three components arise from the non-linear oscillatory nature of the thermoregulatory, blood pressure and respiratory control systems.

The question of interest was therefore whether or not heart rate could be entrained in a similar manner to blood flow. If thermal entrainment of heart rate were possible then this would provide strong evidence of the existence of the thermal component in the RR interval spectrum. Unfortunately, such analysis relies on the RR interval spectrum being expressed in terms of true frequency rather than in beats per interval which is the abscissa scale which results from Fourier transformation of the interval tachogram, i.e. interval plotted against interval number. The raw RR interval wave-form comprises irregular samples and these must be converted to the low pass filtered event series (LPFES) prior to spectral analysis. When the ECG signals recorded during the original entrainment experiments were analysed in this way, the power spectra showed evidence of entrainment in the range 5 to 12 seconds period. What should be borne in mind at this stage is that the range of entrainment of 5 to 12 seconds period is based on a series of experiments in which the two tank temperatures were always 18°C and 46°C. Further work with the entrainment of digit blood flow showed that the range of entrainment was a function of both stimulus frequency and amplitude. A further series of experiments were therefore undertaken in which both stimulus amplitude and frequency were varied incrementally. The results of these experiments confirmed the existence of entrainment in the thermal component of the RR

Figure 3.6 (a) Spontaneous variations in blood flow recorded from the human finger.

(b) Variations in blood flow recorded from the same finger as (a), during the application of a 20-second period thermal stimulus (18°C and 46°C).

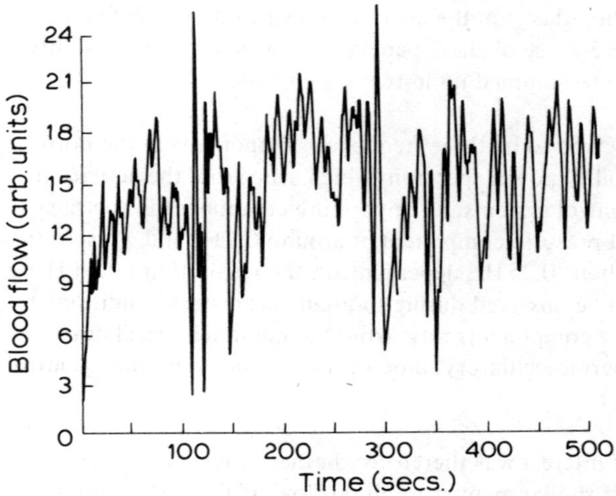

(a)

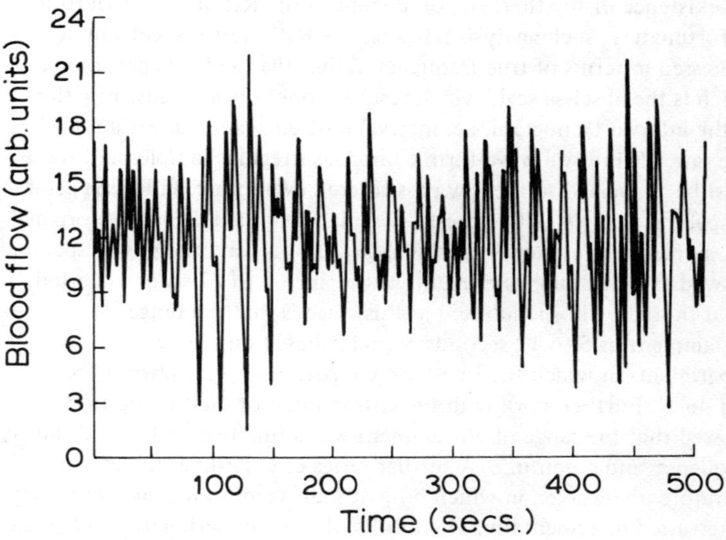

(b)

Figure 3.7 (a) and (b) The power spectrum of wave-forms 3.6 (a) and 3.6 (b), respectively.

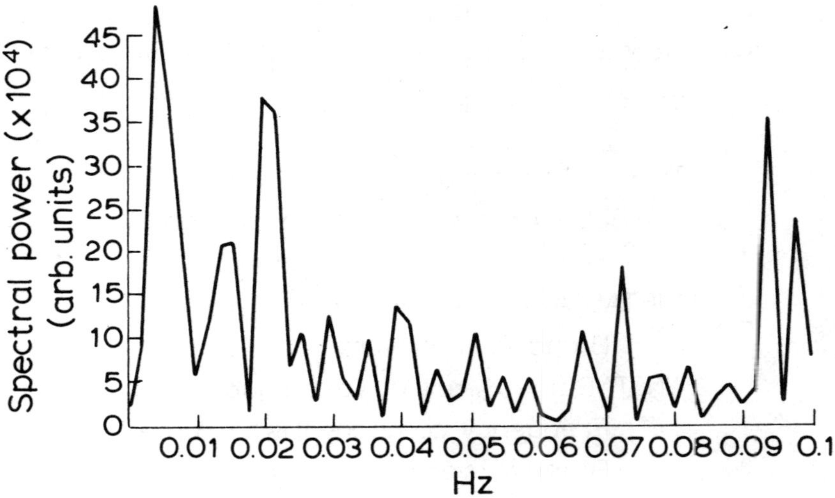

(a)

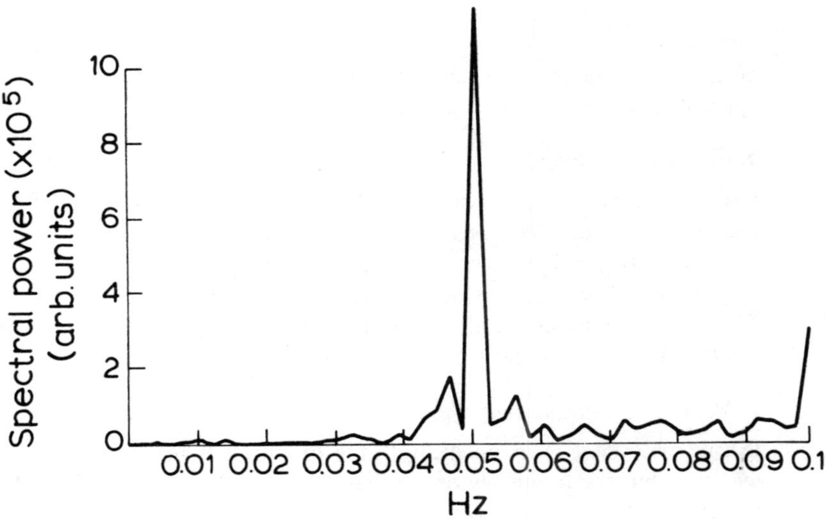

(b)

Figure 3.8 The various types of entrainment found in a biological system.

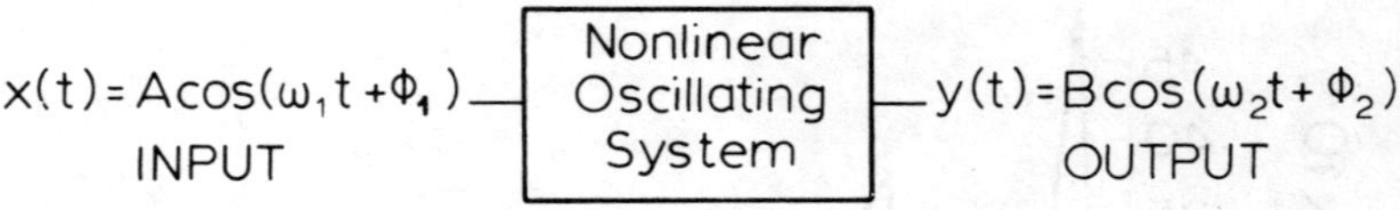

Types of Entrainment:

1. STABLE - Amplitude and Phase stable
   $\omega_2 = \omega_1$, B and Φ_2 constant.
2. METASTABLE - $\omega_2 = \omega_1$ (constant)
   B and Φ_2 may vary.
3. UNSTABLE - ω_2 not always equal to ω_1
   Frequency component sometimes lost,
   B and Φ_2 very unstable.

interval wave-form, but they also showed that there are three types of entrainment: stable, metastable and unstable entrainment. These three types will now be described in relation to Figure 3.8.

1. Stable Entrainment

Referring to Figure 3.8, if an input $x(t) = A \cos(\omega_1 t + \psi_1)$ is applied to the non-linear oscillatory system and the output is $y(t) = B \cos(\omega_2 + \psi_2)$, then for stable entrainment both the amplitude and phase of the output, i.e. B and ψ_2, are constant. (The thermal component of the LPFES power spectrum will be stable in both amplitude and frequency, such that $\omega_2 = \omega_1$).

2. Metastable Entrainment

During metastable entrainment both the amplitude B and the phase ψ_2 may vary but the frequency of the thermal component remains locked to the stimulus frequency ($\omega_2 = \omega_1$).

3. Unstable Entrainment

During unstable entrainment, the output frequency ω_2 is not always equal to the input frequency ω_1, the amplitude and phase of the thermal component of the RR interval wave-form (B and ψ_2 are very unstable and there may be periods when the component at the stimulus frequency (in the LPFES power spectrum) is lost.

The Analysis of Metastable Entrainment

To recapitulate, metastable entrainment of heart variability is categorised by a component in the LPFES spectrum whose frequency is constant and at the stimulus frequency, but whose amplitude and phase may vary in relation to the stimulus. Figure 3.9 shows an example of the phenomenon for a 10-second stimulus with tank temperatures of 22°C and 37°C.

The figure illustrates a 512 point running LPFES power spectrum computed from a 256-second data window and moving along 50 samples at a time. The instability in power can be clearly seen. In fact the unstable amplitude seems normally to result from the unstable phase. HRV signal shows the effect (Figure 3.10). Referring to Figure 3.10a, the figure illustrates the type of perspective plot which is obtained with a completely stable thermal component. (The plots of Figures 3.10a and 3.10b are constructed by dividing a length of HRV wave-form into two stimulus period segments with time increasing from front to back.) The effect of metastability can clearly be seen by comparing the two figures. In the metastable case (b) both the amplitude and phase of the wave-form are varying.

The most effective means of studying the phenomenon of metastability seems to be the running phase spectrum. In this application, the phase of the first harmonic (stimulus frequency) is plotted as a function of time. Stable entrainment produces a flat phase spectrum while for metastability the flat regions are interspersed with areas of unstable phase.[8]

The transition from metastability to full stability can also clearly be seen by computing the power in the bands from the 512-point running LPFES power spectra, as shown in Figure 3.11.

Summary of Metastable Entrainment

In summary, therefore, metastable entrainment differs from stable entrainment by virtue of the fact that the amplitude and phase of the stimulus component in the HRV power spectrum varies. Our results

Figure 3.9 An example of metastable entrainment with a 10-second thermal stimulus.

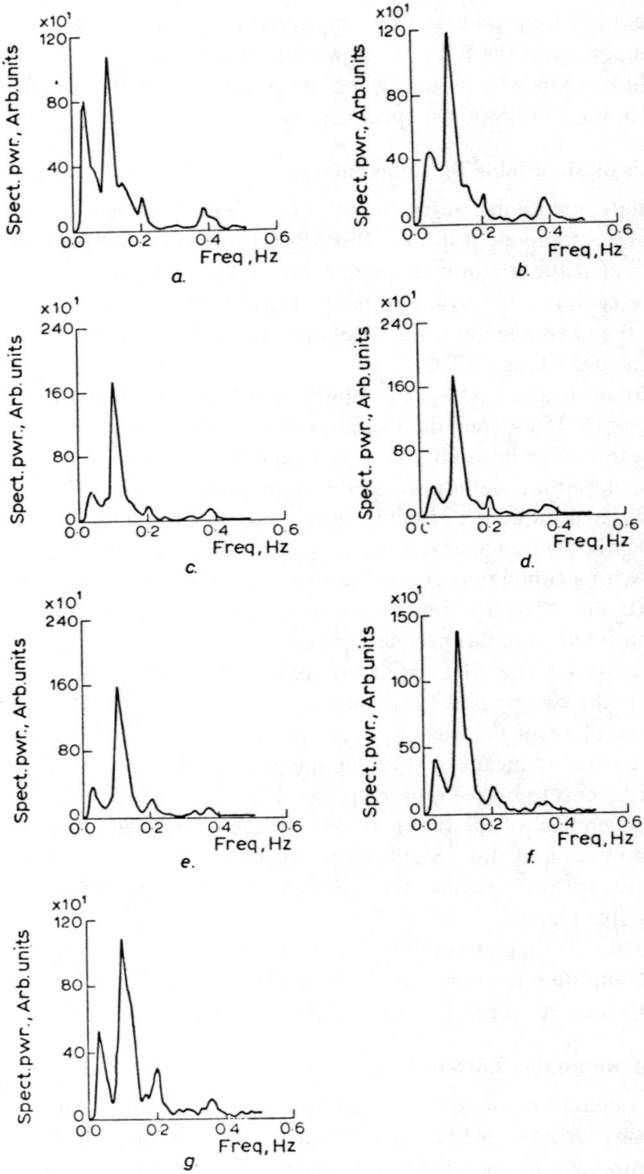

Figure 3.10 Examples of stable (10a) and metastable (10b) entrainment using perspective plots.

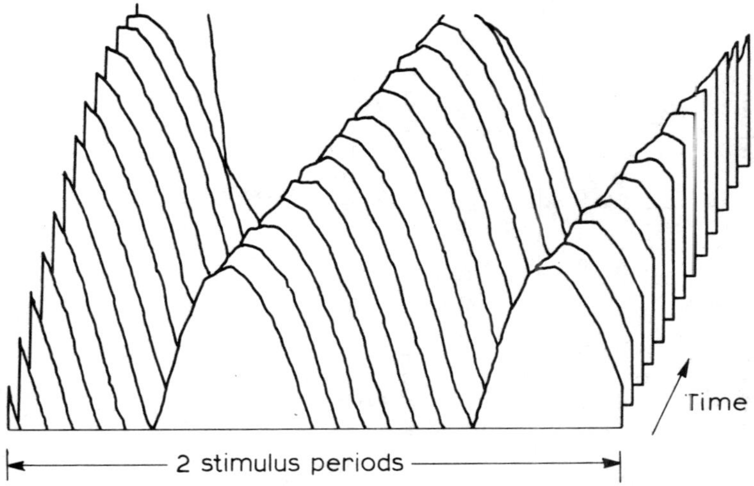

(a)

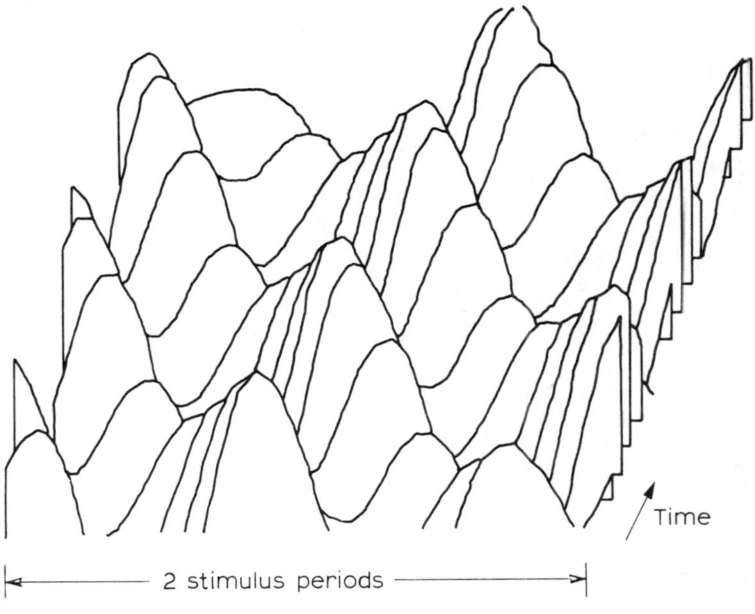

(b)

Figure 3.11 Power distribution in various frequency bands of the LPFES showing a transition from metastable to stable entrainment.

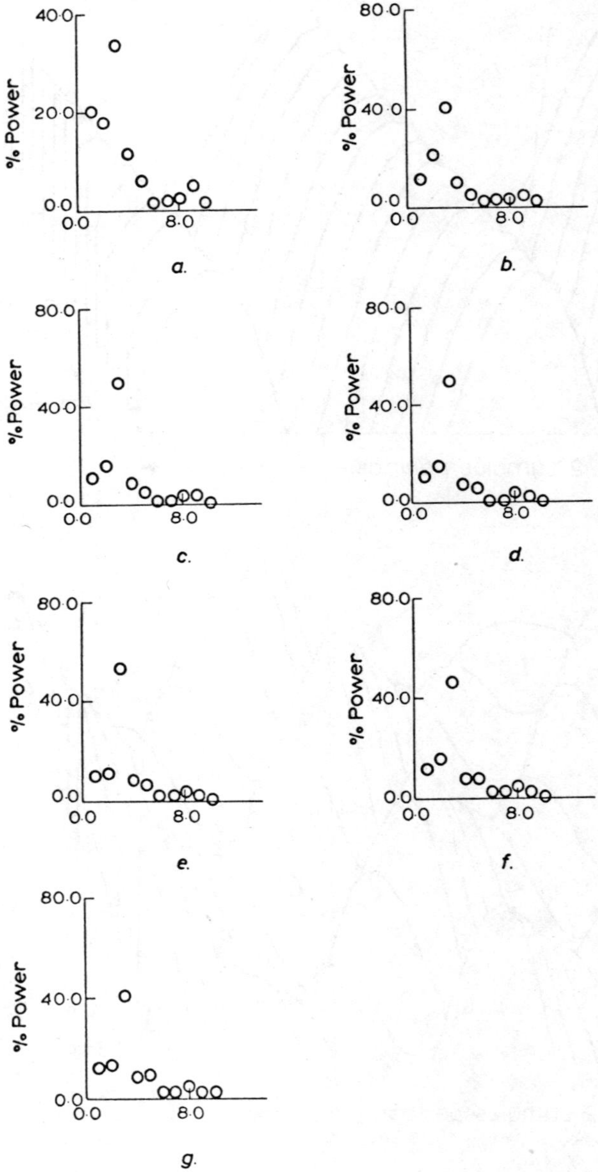

indicate that it is the phase which is the important factor, particularly when one is observing the transition from metastability to full stability.

Unstable Entrainment

The condition of unstable entrainment is categorised (referring to Figure 3.6) by a combination of the following conditions: (1) ω_2 is not always equal to ω_1, (2) the stimulus frequency component sometimes disappears from the LPFES power spectrum, and (3) B and ψ_2 are very unstable. Figure 3.12 illustrates this phenomenon. Referring to Figure 3.12a, even though a thermal stimulus is present throughout the record, there is no evidence of it in the power spectrum. However, if the LPFES wave-form is examined more closely, it becomes clear that there are sections of the wave-form where there is an oscillation at the stimulus frequency – this is an example of unstable entrainment.

Hence the phenomenon of unstable entrainment occurs in the presence of a constant thermal stimulus. The exact reason why the wave-form switches from one state to another at a particular time is often obscure. Nevertheless, unstable entrainment is frequently caused by either bursts of biological noise or non-linear beating between the stimulus and other physiological oscillations which form the HRV wave-form.

The Relationship between Metastable and Unstable Entrainment

A phenomenon which is often observed in the experimental data is that of the HRV wave-form oscillating between metastable and unstable entrainment. In the introductory section the three types of entrainment (Figure 3.6) were discussed in relation to a spontaneously oscillating non-linear system of natural frequency ω_2. In practice, the frequency of the output, in this case the thermal component of the HRV wave-form, is not entirely stable and it is this instability which provides a plausible explanation for the transition phenomenon.

Figure 3.1 illustrates the relationship between stimulus frequency and natural frequency. Let us assume that the applied stimulus is of

frequency f_{STIM_1} and $f_{NAT_2} \leqslant f_{NAT} \leqslant f_{NAT_1}$.

For these conditions the stimulus frequency always lies within the range of entrainment (the sloping section of the characteristic) and the system always resides in the metastable state. If, however, the stimulus frequency is f_{STIM_2} then it can be seen from the diagram that as the natural frequency of oscillation shifts from f_{NAT_2} to f_{NAT_1}, a point

Figure 3.12 An example of unstable entrainment (6-sec. thermal stimulus) where the power spectrum shows no evidence of the stimulus while regions of the LPFES do.

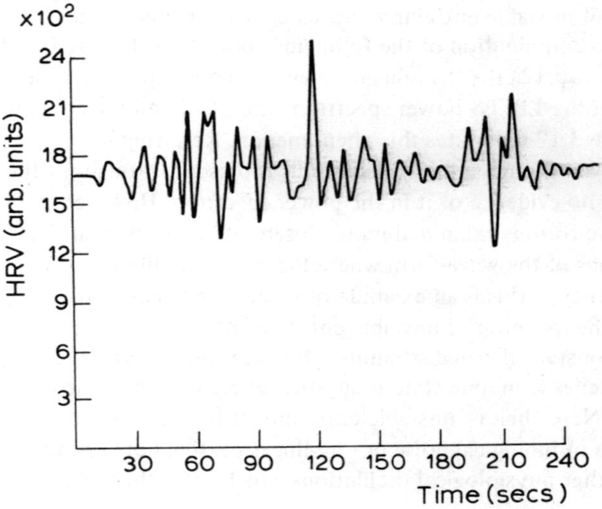

(a)

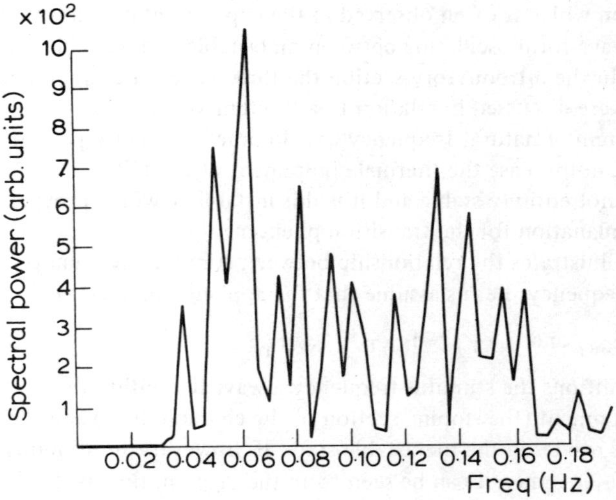

(b)

Figure 3.13 The relationship between stimulus frequency and the natural frequency of oscillation of a biological system.

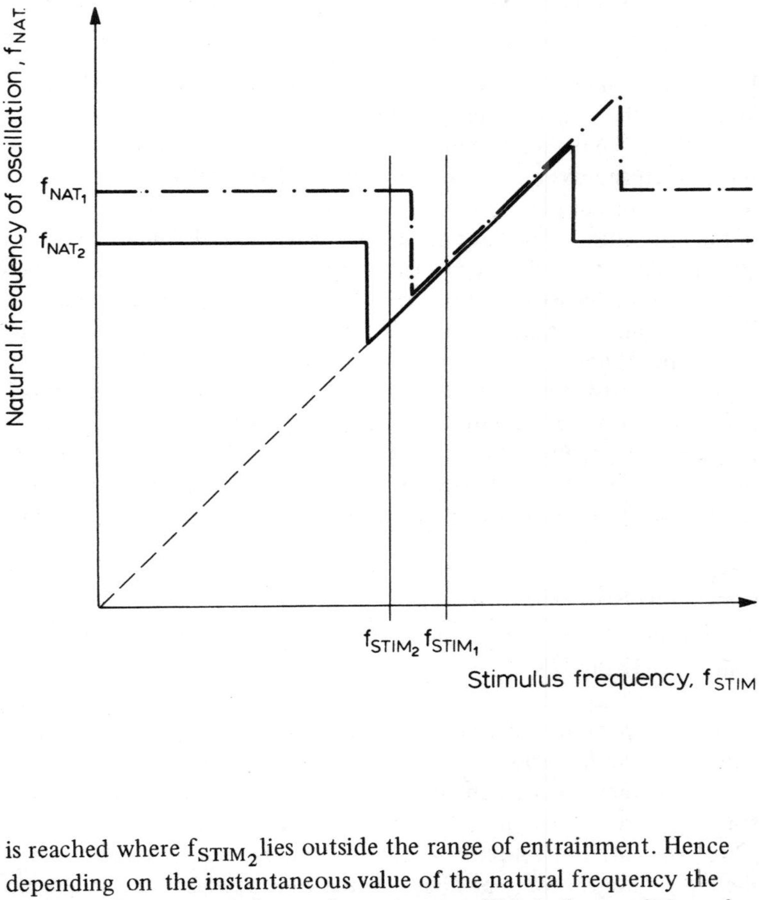

is reached where f_{STIM_2} lies outside the range of entrainment. Hence depending on the instantaneous value of the natural frequency the system will move in and out of entrainment. This is the condition of unstable entrainment.

The interesting question, of course, is what is the physiological origin of the instability in the natural frequency of oscillation? In an earlier paper[6], the basic structure of the non-linear system was described as comprising a negative feedback regulatory system whose efferent pathway comprised a non-linear Bang-Bang element and a linear filter incorporating pure time delay. The linear filter element and pure time delay in the context of our work represent the characteristics of the peripheral vascular bed (the peripheral resistance) G(s) first defined by Scher and Young,[9] where

$$G(j\omega) = \frac{22.0\,(1 + 12\,j\omega)}{(1 + 92\,j\omega)\,(1 + 2\,j\omega)}.$$

Heart rate variability has been studied[10] using this model in order to investigate the origin of the variability in the natural frequency of oscillation. The parameters of the model were systematically varied, i.e. the time constants and the pure time delay. The results showed that wide variations in the time constants produced only a minor difference in the frequency of oscillation, whereas even relatively minor changes in the time delay caused large changes in the frequency of oscillation. Our conclusion therefore is that the variability in the frequency of the thermal component arises from fluctuations in the pure time delay. In their paper Scher and Young describe the response of the smooth muscle comprising the peripheral resistance by a first order differential equation incorporating pure time delay. On this basis, therefore, the physiological origin of the fluctuations in the natural frequency of the thermal component of the HRV wave-form would seem to be variations in the time delay associated with the response of the peripheral smooth muscle.

Concluding Remarks

The results of the investigation which have been described can be summarised as follows:

(a) The spontaneous heart rate variability wave-form has a thermal component in the region of 0.05 Hz.

(b) The thermal component can be entrained by a periodic thermal stimulus of suitable amplitude and frequency.

(c) The entrainment phenomenon for the thermal component of the HRV wave-form can be divided into three types, stable, metastable and unstable entrainment.

(d) Continuous movement of the system between metastable and unstable entrainment seems to result from fluctuations in the natural frequency.

(e) The physiological origin of these fluctuations may be due to changes in a pure time delay associated with the characteristics of the peripheral vascular smooth muscle.

In the wider context many of the ideas and methods which have been developed here can be used in teaching a multi-disciplinary approach to the analysis of physiological systems. For example, in the material which has been presented, there are strong elements of physiology, control theory and signal analysis. If treated in isolation it is often very difficult to convince the student of the relevance of a particular technique or approach. Hence both motivation and retention are often extremely low. However, when such techniques are introduced within the framework of a physiological investigation their relevance is immediately apparent and consequently motivation is often extremely high.

References

1. Burton and Taylor, 'A Study of the Adjustment of Peripheral Vascular Tone to the Requirements of the Regulation of Body Temperature', *American Journal of Physiology*, 129 (1940) 565-77.

2. R.I. Kitney, 'The Thermoregulatory Control System in Man and its Study by Digital Computer Simulation', PhD thesis (London, 1972).

3. N. Minorsky, *Non-linear Oscillations* (Van Nostrand, London and New York, 1962).

4. R.I. Kitney, 'An Analysis of the Non-linear Behaviour of the Human Vasomotor Control System', *Journal of Theoretical Biology*, 52 (1975) 231-48.

5. J.S. Bendat and A.G. Piersol, *Random Data: Analysis and Measurement Procedures* (Wiley Interscience, New York and London, 1971).

6. B.W. Hyndman, R.I. Kitney and B. McA. Sayers, 'Spontaneous Rhythms in Physiological Control Systems', *Nature*, 233, 5318 (1 October 1971) 339-41.

7. B. McA. Sayers, 'Analysis of Heart Rate Variability', *Ergonomics*, 16, 1 (1973) 17-32.

8. R.I. Kitney and O. Rompelman, 'Some Aspects of Heart Rate Variability', in *Biomedical Computing*, ed. J. Perkins (Pitman Medical, London, May 1977).

9. A.M. Scher and A.C. Young, 'Servoanalysis of the Carotid Sinus Reflex on Peripheral Resistance', *Circulation Research*, 12 (1963) 152-62.

10. R.I. Kitney, O. Rompelman and A. Spruyt, 'The Analysis of Stable and Unstable Oscillations in Heart Rate Variability', *Proceedings of the Eleventh International Conference on Medical and Biological Engineering* (Ottawa, 1976).

4 MODELLING IN THE TEACHING OF MATHEMATICS

4.1 Introduction

The construction of a series of mathematical relationships to describe, albeit approximately, the behaviour of a certain physical system is known to scientists as modelling and in recent years has become increasingly important and easy as availability of computer systems has enabled the predictions of such models to be investigated. In many cases this model building activity has involved mathematicians, but their role has changed recently as numerical rather than analytic solutions are sought with the aid of computers. This activity involving computers clearly has an important role in undergraduate mathematical education.

4.2 Some Uses of Population Models in the Teaching of Mathematics

B.W. Conolly and F. Oliveira-Pinto

In this section we discuss a family of mathematical models with applications in physical and sociological as well as biological research. They are intuitively appealing as well as relevant while the range of methodology used in their analysis contains a large fraction of the mathematical apparatus needed by practitioners who operate in the world of so-called 'real life' problems. At its simplest the analysis requires little in the way of mathematical sophistication, but the deeper penetration required for, say, planning purposes, may necessitate the deployment of advanced theoretical methods and, often, the interaction between computer simulation and orthodox analysis can be illuminating and even, in some cases, essential.

The models with which we are concerned describe the fluctuations in time of a physical or biological 'mass' arising from a variety of physical or other causes. In the biological context these models are known as *population models*, although their application is certainly not so narrowly confined as this description suggests. Their analysis may be either *deterministic*, in which case the formulation usually involves quite simple differential or difference equations: or, alternatively the analysis may have a probabilistic basis and in this case the methodology is that of the theory of stochastic processes. A great deal of information about the dependence of the fluctuations can often be derived by the simplest of mathematical methods. For planning purposes and deeper insight a probabilistic treatment is usually more relevant. The interface between the two approaches makes an important and instructive topic.

To illustrate the simplest case we consider a biological mass of size N(t) at time t. In the most common application N(t) will be a population size measured in numbers of individuals (fish, insects, plants, birds, germs. . .). Strictly, N(t) is then a non-negative integer but, as we shall see, it is convenient to model it as a continuous non-negative real number and in this case one can think of N(t) as representing a kind of average population size. It is this interpretation which it is instructive to compare with a genuine average obtained from a suitable stochastic model. The variation with time of the population depends on such easily grasped entities as the birth rate and the death rate. To illustrate another source of population change we shall later examine briefly the effect of hunting. Both birth and death rates depend strictly on the age structure of the population but in the simplest case they can be expressed as time independent *averages* over the whole population. Thus we may denote by a the mean number of new individuals produced in unit time by one member of the population, and by b the mean number of deaths in unit time per member of the population. This leads to the first order differential equation

$$\frac{dN}{dt} = [a - b]\, N \tag{1}$$

(we do not always write N(t) when the dependence is evident) whose solution is

$$N(t) = N_0 \exp(a-b)t, \tag{2}$$

where $N_0 = N(O)$, and

$$\epsilon = a - b \tag{3}$$

can be described as the average net birth rate. When ϵ is positive, (1) is a (favourable) approximation to the growth of money invested at continuously compounded interest. In this context it is instructive for educational purposes to contrast the *continuous time* solution (2) with the *discrete time* solution

$$N_n = N_0\, [1 + \epsilon]^n \tag{4}$$

to the *difference equation*

$$N_n = N_{n-1}[1 + \epsilon], \tag{5}$$

where N_n is the amount of capital when interest has been compounded n times, and ϵ is the interest rate. There are educational advantages in comparing the formulations, the methods of solution and the numerical behaviour of the solutions. Another obvious application of (1) is to radioactive growth ($\epsilon > 0$), and decay ($\epsilon < 0$).

To the model just described there are two common stochastic counterparts. Both of these can be analysed completely in closed form, providing an excellent illustration of the mathematical techniques required. The two stochastic approaches themselves illustrate examples of modelling each of which is demanded uniquely by certain physical situations but which in the present case represent useful *alternative* methods of approach to the analysis. The first approach supposes that changes of ±1 at discrete but random intervals are the only ones to occur. The second approach conceives of very small continuous changes all the time. This latter, which we have no space to describe here, leads to the formulation of differential equations of *diffusion* type for the probability density of population size at time t; the former leads to the formulation of a system of infinitely many differential difference equations for the probabilities $p_n(t)$ that the random population size N(t) consists of n individuals at time t. Under suitable circumstances the passage from one model to the other can be made and provides another excellent educational illustration of the relationship between two apparently quite widely contrasted methods of approach. Taking the approach in which N(t) becomes a random variable (stochastic process), integer valued and non-negative, we repeat that we seek to calculate

$$p_n(t) = P(N(t) = n \mid N(O) = N_0) \tag{6}$$

The quantities a and b occurring in the deterministic formulation are now interpreted *consistently* with that formulation by the assertion that anh + o(h) and bnh + o(h) are the transition probabilities from N(t) = n to n + 1, and from n to n–1, respectively, over a small but finite time interval (t, t + h). The consistency of the interpretation resides in the fact that in the deterministic case we claim that on average anh new individuals are added to the population during the interval (t, t + h). By making h sufficiently small the average number anh can be interpreted as the probability of a birth multiplied by 1, so that numerically anh has the characteristic of a probability. Establishment of the differential difference equations satisfied by

$p_n(t)$ is one of the most frequently encountered techniques in the analysis of integer valued stochastic processes. It is important to notice the characteristic of the model that the probability of no change in a small interval (t, t + h) is 1–(a + b)nh + o(h) when the population size is n; this is the dominant probability. By standard arguments we obtain for the set $p_n(t)$: n → 0, 1, 2. . .the infinite collection

$$\frac{d}{dt}p_0(t) = bp_1(t),$$

$$\frac{d}{dt}p_1(t) + [a + b]\; p_1(t) = 2bp_2(t),$$

$$\frac{d}{dt}p_n(t) + [a + b]\,np_n(t) = [n + 1]\,bp_{n+1}(t) + [n-1)\,ap_{n-1}(t),$$

$$(n \geqslant 2). \tag{7}$$

The method of generating functions can be effectively used in this case and accordingly we introduce

$$P(x, t) = \sum_{n \geqslant O} p_n(t)\, x^n, \quad (O \leqslant x \leqslant 1). \tag{8}$$

Then P (x, t) satisfies

$$\frac{\partial P(x, t)}{\partial t} + [ax-b]\,[1-x]\,\frac{\partial P(x, t)}{\partial x} = O, \tag{9}$$

and hence

$$P(x, t) = \left[\frac{b[1-e^{-\epsilon t}]-x[b-ae^{-\epsilon t}]}{a-be^{-\epsilon t} - ax[1-e^{-\epsilon t}]}\right]^{N_o}, \tag{10}$$

having taken $P(x, O) = x^{N_0}$.

This is an elementary but instructive exercise. To show the great power of generating function methods it is convenient to introduce a moment generating function from which differentiation produces the moments directly. Thus for the mean N and variance Var(N) the expressions

$$\hat{N} = N_0 e^{\epsilon t}, \tag{11}$$

and

$$Var(N) = \frac{\hat{N}^2}{N_0}[1-e^{-\epsilon t}]\,[\frac{a + b}{\epsilon}] \tag{12}$$

are obtained.

Equation (10) contains complete information about the set $p_n(t)$ and it is comparatively straightforward to extract the coefficients of x^n in order to specify $p_n(t)$. The simplest is

$$P_0(t) = \left[\frac{b[1-e^{-\epsilon t}]}{a-be^{-\epsilon t}}\right]^{N_0}. \tag{13}$$

Comments which may be made and illustrated in the context of the example are as follows:

1. Comparison of $\hat{N}$ with the Deterministic Solution

There is no difference and in this case the simple deterministic approach yields the same conclusion about the mean as the rather more elaborate stochastic analysis. It should not be concluded that this is always the case. Moreover our knowledge about the behaviour of the population must always be regarded as inadequate until we know at least something about the *dispersion* of the population about the mean. This is measured by Var(N). The most striking observation arises from the case when $a = b$ and $\epsilon = O$. It might have been expected that when births balance deaths the population is in a stable condition but Equation (12) shows this to be not at all the case. Instead we have

$$\text{Var}(N) = 2N^2bt/N_0 \tag{14}$$

so that the distribution spreads increasingly about the mean as $t \to \infty$. The fact that σ_N is proportional to $t^{1/2}$ is itself worthy of comment and is indeed consistent with the behaviour of a large class of stochastic processes which one might expect to be stable under conditions of equal average input and output. An instructive example is provided by the model of a bank account and can stimulate much surprise as well as sighs of comprehension.

2. Probability of Extinction of Population

It is important to biologists to be able to make probability statements about the survival prospects of a population under given birth and death (and other) characteristics. Only a stochastic model can help here. When $a > b$ ($\epsilon > O$) the deterministic model merely predicts an exponential increase but we know that at every time t there is a finite probability nevertheless that the population has already become extinct. This is given by $P_0(t)$ and, irrespectively of whether $a \geqslant b$ or $a \leqslant b$, we have

$$Po(t) = [b(1-e^{-\epsilon t})/(a-be^{-\epsilon t})]^{N_0}. \tag{15}$$

When $b < a$ this tends to $\left[\frac{b}{a}\right]^{N_0}$ with increasing t, and in turn tends to zero as $N_0 \to \infty$. All $p_n(t)$ for $n \neq O$ tend to zero *and so there is a non-zero probability* $1-\left[\frac{b}{a}\right]^{N_0}$ *that the population becomes infinitely* large. On the other hand, when $b > a$, $p_0(t)$ must be replaced by unity as $t \to \infty$ for all N_0. The important fact is that the scale of the approach to the limit is indicated.

3. Computer Simulation

There is no reason to utilise computer methods other than for routine calculation in a simple case like this. But for the purposes of illustration of procedure and to provide the opportunity to acquire familiarity and competence a simple model whose behaviour is completely known provides an invaluable training ground.

The Pearl-Verhulst (Logistic) Model: The Effect of Hunting

We show how by a simple extension of the model the objection can be overcome that when $a > b$ the population cannot possibly grow indefinitely simply because the resources of the environment, necessary for its survival, are in practice finite. The modification associated with the names Pearl[1] and Verhulst consists in reducing the effective birth rate ϵ by an amount proportional to population size. The result is the differential equation

$$\frac{dN}{dt} = [\epsilon - \gamma N]\, N \qquad (16)$$

which gives the solution

$$N(t) = \frac{K}{1 + Ce^{-\epsilon t}}. \qquad (17)$$

γ is a positive constant which may be interpreted as an indicator of the greed of the population with respect to the food supplied by the environment. The solution shows an increase to the steady value $N(\infty) = K = \frac{\epsilon}{\gamma}$, provided that $\epsilon > O$ and that $N_0 < K$ (so that there is an initial increase in the population). C is the integration constant. The alternative name given to the solution (17) is *Logistic Law.* An interesting model in which it arises naturally is the very simplest epidemic model 'without removal'. A population has a fixed size N_0. At any time it may be divided into x(t) infectives (carriers of such a mild disease as the common cold) and N_{0-x} susceptibles (members who will catch the disease upon contact with an infective). The disease is spread by homogeneous mixing at a rate proportional to xy, that is, infectives increase (as they catch the disease and are therefore, because not removed to isolation, capable of spreading it further) at a rate say $cx[N_0-x]$, where c is constant. This gives the logistic law too and the obvious behaviour of dx/dt as a function of t, rising to a maximum when half the population is infected, is a phenomenon for which there is supporting evidence. Another less morbid interpretation would be to the spread of rumours. In that case x would be the carriers of a rumour, telling it to the uninitiated part of the population by a homogeneous mixing mechanism! A stochastic counterpart to the Logistic Model (with transition infinitesimal probabilities $a-\gamma N$ and b per population member) shows that the statistical average is LESS than that given by the deterministic model.[2]

It is instructive to modify Equation (16) for other purposes. One simple extension to (16) allows one to investigate the effects of hunting, and by this we have in mind the shooting in some countries of wild birds. This topic is treated usually on the basis of emotion so that a more logical approach would appear entirely appropriate. Crudely, hunting is permitted only in the non-breeding season. As a first approximation one can divide the year into two equal periods of duration ½T each. Then during the breeding season the population is deterministically governed by

$$\frac{dN}{dt} = [a-b]N-\gamma N^2, \qquad (O \leqslant t \leqslant \tfrac{1}{2}T), \tag{18}$$

and otherwise by

$$\frac{dN}{dt} = -bN - \gamma N^2 - H, \quad (\tfrac{1}{2}T \leqslant t \leqslant T), \tag{19}$$

where H is a constant, the average number of birds of the species under consideration permitted to be destroyed by hunting per time unit. Given N_0 one can then determine N(½T), which is the initial value when the hunting season begins and Equation (19) takes over. From this one can calculate N(T), the value of the bird population at the end of the year. There will be a balance if $N(T) = N_0$ and accordingly the permissible level H of hunting could be calculated in order that this is the case. But for planning purposes a stochastic model is needed. We suspect, moreover, that it would be prudent to arrange that N(T) is somewhat larger than N_0 in order to avoid the fluctuation phenomenon noted earlier. Since stochastic analysis, the details of which we shall not embark upon here, is likely to be somewhat more difficult than in the simple birth and death model. We propose the model as a challenging and environmentally meaningful subject for computer simulation. An alternative model would consider the rate of removal of individuals by hunting as proportional to the population size. In this case Equation (19) would be replaced by $\frac{dN}{dt} = -(b + h)N - \gamma N^2$.

Other Applications of Single Population Models

A more general single population model can be represented deterministically by the equation

$$\frac{dN}{dt} = [a-b]N + c - d. \tag{20}$$

where a, b, c and d are all positive. a and b continue to play the role of birth and death rates per unit of population. c, however, measures the rate of increase in the population due to *immigration*, and d that due to *emigration.* These phenomena are independent of population size. By putting a = b = O we have a simple *queueing* model where c is the mean number of demands made on the system per unit time and d is the mean service rate. There is a vast literature devoted to the stochastic aspects of service models, but the fact remains that Equation (20) is a deterministic version of the simplest model of queue size, which would be N(t) – (average number of customers being served at time t). And if population models are the parents of queueing theoretic models then they are the uncles and cousins of the vast array of models allied to queueing processes: inventory, reliability, risk, reservoirs, for example. The mathematical methods applicable to the analysis of one model are clearly available for many more.

Several Interacting Populations: The Work of Volterra and Others

An immediate generalisation of the single population model is that in which several populations compete for resources. Deterministically we now have to deal with systems of differential equations and to achieve a comparable stochastic treatment is in general complicated. It seems reasonable to assert that the mathematical methodology available for the analysis of such more general models is relatively limited. As an example we cite the predator-prey model treated by Lotka[3] and with extensions by Volterra.[4] There are now two populations $N_1(t)$ and $N_2(t)$. The former may be regarded as a species which can flourish in the environment without the need for additional resources. Without the logistic term limiting indefinite growth its fluctuation would then be governed by the first order differential equation $d/(dt)N_1\ (t) = \epsilon_1 N_1(t)$. But in the same ecospace there co-exists a second species whose size at time is $N_2(t)$. This, alone, could not survive on the resources of the environment and would then be governed by $d/(dt)N_2(t) = -\epsilon_2 N_2(t)$ ϵ_1 and ϵ_2 are both positive. The interaction is such that the second population, the predators, feed on the first, the prey. This is modelled by reducing the mean birth rate of the prey by an amount proportional to the size of the predator population, and by increasing the negative mean birth rate of the predators by an amount proportional to the size of the prey population. The model is then represented by the pair of simultaneous non-linear differential equations of the first order:

$$\frac{dN_1}{dt} = [\epsilon_1 - \gamma_1 N_1]\ N_1, \quad \frac{dN_2}{dt} = [-\epsilon_2 + \gamma_2 N_1]\ N_2. \tag{21}$$

The four *constants* $\epsilon_1, \epsilon_2, \gamma_1, \gamma_2$ are all essentially positive in this application though we shall remark later on a model where some might vanish. Volterra shows, using an analysis of exemplary simplicity, that N_1 and N_2 vary periodically about a steady state given by putting the derivatives in Equation (21) equal to zero. Let $K_1 = (\epsilon_2)/(\gamma_2)$, $K_2 = (\epsilon_1)/(\gamma_1)$. It is convenient to suppose that $N_1(O) < K_1, N_2(O) < K_2$. The steady state is $N_1 = K_1, N_2 = K_2$ and to examine possible small oscillations about this Volterra writes $N_1(t) = K_1\ [1 + \nu_1(t)]$, $N_2(t) = K_2\ [1 + \nu_2(t)]$ obtaining

$$\frac{d\nu_1}{dt} = -\epsilon_1 \nu_2\ [1 + \nu_1], \quad \frac{d\nu_2}{dt} = \epsilon_2 \nu_1\ [1 + \nu_2]. \tag{22}$$

Using a technique standard in mechanics we neglect second order terms and then find sinusoidal oscillations with bounded amplitude and frequency proportional to $[\epsilon_1 \epsilon_2]^{½}$. This was hardly to be guessed. On the basis of this model Volterra enunciated three celebrated laws, some of which he claims were supported by contemporary fisheries data. An interesting exercise in methodology would be to calculate the deviation from the true solution caused by the approximation. Volterra published nothing on these lines and indeed concentration on non-linear phenomena belongs to a later period. Such an exercise is of limited interest from the practical point of view because of the crude nature of the model itself.

Stochastic Analysis: Lanchester's Laws

The labour of creating a corresponding analysis increases very heavily with the number of populations in question. We examine briefly a special case of Equation (21) which was proposed in an entirely different context by Lanchester.[5] In this case $\epsilon_1 = \epsilon_2 = O$, and $\gamma_2 < O$. Neither population has births and in a succession of 'encounters' both populations erode each other until one of them becomes extinct. This is Lanchester's so-called Linear Model and it was suggested as a mathematical representation of warfare in ancient times when two armies were drawn up in opposition and each side fired on the other, without being able to take individual aim. In this case the average number of enemy units destroyed by each member of a given 'side' is proportional to the size of the enemy force at that time. The equations corresponding to (21) are:

$$\frac{dN_1}{dt} = -\gamma_1 N_1 N_2, \quad \frac{dN_2}{dt} = -\gamma_2 N_1 N_2, \tag{23}$$

where $\gamma_1 > O, \gamma_2 > O$. (We have changed the sign of γ_2 in (21).) γ_1 and γ_2 are clearly in some sense measures of each side's military effectiveness. They are commonly known as *attrition coefficients*. The obvious deduction from (23) is

$$\gamma_2 N_1 - \gamma_1 N_2 = \gamma_2 N_1(O) - \gamma_1 N_2(O) \tag{24}$$

which means, deterministically, that if the initial strengths are such that the right-hand side of (24) is positive then also the left-hand side remains positive at least until N_2 vanishes when the model ceases to be meaningful. But this means that the side with subscript 1 is the victor since he still has troops when the side with subscript 2 has been annihilated. The linear nature of (24) gives the model its name. A complete solution of the equations is an elementary exercise.

It is of much greater interest to examine the stochastic counterpart of the model and in particular to attempt to calculate the probability that N_1, the force with subscript 1, is the victor. Let $u(N_1, N_2)$ be the conditional probability that N_1 is eventually the victor, given that the opposing strengths are N_1 and N_2 units respectively. Following the argument used to calculate (6) we suppose that at an encounter, irrespective of when it takes place, either N_1 or his adversary N_2 loses a single unit. Then if $p_{N_1 N_2}$ is the probability, given state (N_1, N_2) of the the two sides, that N_2 loses one unit, to be consistent with (23) requires that

$$p_{N_1 N_2} = K\gamma_2 N_1 N_2 \text{ and } q_{N_1 N_2} = K\gamma_1 N_1 N_2,$$

where $q_{N_1N_2} = 1 - p_{N_1N_2}$ is the probability of the alternative outcome of the encounter, namely, that N_2 loses a unit. K is chosen so that the probabilities sum to unity and thus we find that

$$p_{N_1N_2} = \frac{\gamma_2}{\gamma_1 + \gamma_2}, \qquad q_{N_1N_2} = \frac{\gamma_1}{\gamma_1 + \gamma_2} \tag{25}$$

which, conveniently independent of N_1 and N_2, permits us to drop the subscripts on p and q. At the first encounter either N_1 or N_2 loses a unit and the duel starts, as it were, afresh. Then denoting initial strengths by N_{10} and N_{20} we have

$$u(N_{10}, N_{20} = pu(N_{10}, N_{20} - 1) + qu(N_{10}-1, N_{20}) \tag{26}$$

which is a *partial difference equation* and a very different animal from anything so far encountered. For consistency the boundary conditions $u(O, N_{20}) = O$, $u(N_{10}, O) = 1$, $(N_{10} \geqslant 1, N_{20} \geqslant 1)$ are imposed. The solution to this problem is, however, more easily obtained by the more fundamental observation that under the conditions of the model each encounter is a Bernouilli trial with probability p of success, where success means that N_1 removes one N_2 unit. The extinction of N_2 occurs on the N_{20}th success and so the problem is recognised as elementary and the solution is the routine procedure of writing down the probability that the N_{20}th success occurs at the $(N_{20} + k)$th trial $(k = 0, 1, 2,\ldots, N_{10} - 1)$ and summing over k to obtain $u(N_{10}, N_{20})$. The probability that the N_{20}th success occurs at the $(N_{20} + k)$th trial is simply the product of the probability.

$$\binom{N_{20} + k-1}{N_{20}-1} p^{N_{20}-1} q^{k}.$$

that in the first $N_{20} + k-1$ trials there are $N_{20}-1$ successes, and p, the probability that the last trial is a success. Hence we can write

$$u(N_{10}, N_{20}) = p^{N_{20}} \sum_{k=O}^{N_{10}-1} \binom{N_{20} + k - 1}{N_{20} - 1} q^{k}. \tag{27}$$

It is useful to verify that (27) satisfies (26) and the associated boundary conditions. The methodological lesson to be learned is that to derive a solution *directly* by utilising the fundamentally simple nature of an otherwise apparently difficult problem is easier and more meaningful than to struggle to get a solution indirectly by mechanical means.

To conclude these remarks we point out that this model provides an excellent opportunity to illustrate the Central Limit Theorem, for the number of encounters until extinction of N_2 is the sum of independently and identically distributed geometric variables. The Central Limit approximation to $u(N_{10}, N_{20})$ also yields a *probabilistic* interpretation to the deterministic condition $\gamma_2 N_{10} - \gamma_1 N_{20} > O$ for the victory of N_1 for in this case $u(N_{10}, N_{20})$ tends asymptotically to unity.

Conclusion

The content of this section is not original. It has been drawn from a variety of sources, and we suggest that it will be of great benefit to the interested reader to peruse those papers for himself. Such originality as we have displayed resides in our proposal of what we consider to be a versatile family of mathematical models to an audience with a primary interest in education in mathematical model building and methodology, especially to life science students. Population models are very much in the public eye at the present time because of their relevance to problems of current world interest, environmental, health, sociological. Immense advances have been made in the most recent times which the active educator will find it an advantage to explore.

References

1. R. Pearl, 'The Biology of Death I-VII', reprinted from *The Scientific Monthly* (Lippincott, Philadelphia, 1921).
2. W. Feller, 'Die Grundlagen der Volterraschen des Kampfes ums Dasein in wahrscheinlichkeitstheoretischer Behandlung', *Acta Biotheoretica*, 5 (1939) 11-40.
3. A.J. Lotka, *Elements of Mathematical Biology* (formerly *Elements of Physical Biology*) (Dover, New York, 1957). First published in 1924.
4. V. Volterra, 'Variazioni e fluttuazioni del numero d'individui in specie animali conviventi', *Memorie dell'Accademia dei Lincei* (Roma), 2 (1926), 31-113.
5. F.W. Lanchester, *Aircraft in Warfare: The Dawn of the Fourth Arm* (Constable, London, 1916).

4.3 The Place of the Computer in Teaching Applied Mathematics

R.D. Harding

In most of the modes of use of the computer to assist in teaching or learning, the computer itself is a *medium* for teaching. In the application to be described here, it is used rather as a *tool*.

The Traditional Approach

In the traditional approach to applied mathematics problem solving, three stages can be distinguished:

Formulation Solution Interpretation

Formulation is the process of setting up the problem or system to be studied in a mathematical form. Often this will yield a set of differential equations. In the solution stage, these mathematical equations are manipulated by analytical methods to obtain explicit formulae for the variables of interest. Interpretation is not entirely separate from solution; at this stage the properties of the solution begin to be perceived and related to the principles used in formulating the problem.

If, in the education of mathematicians, one were only concerned with technique, the solution of any particular problem would merely be regarded as an exercise for practising that technique; formulation and interpretation would not be emphasised. However, if one has wider aims – for instance teaching the whole technique of problem solving, or teaching the principles of a subject such as fluid dynamics – then the solution stage, although vital, is only a step on the road to obtaining further insight into the subject being studied.

It is at this point that the weakness of traditional techniques lie. Sometimes they lead quickly to a result whose properties are easily understood, but at other times the process of getting a solution can be long, and in that case tends to distract attention from the study of the subject as a whole. It is also quite likely that no analytic solution exists, and even when it does, small modifications of the problem can lead to cases where it does not.

For example, in elementary dynamical theory, study of the motion of the bob of a pendulum leads to the simple harmonic equation

$$\ddot{X} + w^2 X = 0$$

provided X (the angle of inclination) is small. As soon as motions where X is not always small are considered, the equation is modified to

$$\ddot{X} + w^2 \sin X = 0.$$

The former has a simple analytic solution, the latter does not. Notice that the solution of the first equation is only said to be analytic because it involves familiar functions (sin wt and cos wt). If one wanted numerical values, it would be necessary to consult tables.

The Advantages of Computational Analysis

Computational analysis offers an escape from this dilemma. A well-posed physical problem can always be solved numerically and a relatively small number of numerical methods will cover the majority of problems. For instance, any problem that involves an ordinary differential equation with initial values can almost certainly be solved using the Runge-Kutta method. It would work just as well for either of the two equations discussed above.

The value of this approach is increased all the more if graphical display facilities can be used to display the results. Then the essential properties of the solution are seen at once and the student can begin to interpret his solution. In doing this he is helped by being able to run his program again with a variety of data: the effect of changing parameters can be quickly seen on the display. This latter advantage is to some extent independent of the full use of computational analysis, because it is possible to provide biology students with ready made programs to explore a particular problem and indeed computers are widely used in universities in just this way. However, if this is done, a number of other benefits are lost because the computer is being used as a *medium* rather than a *tool*: the distinction lies in the amount of control the learner can exercise over the resources at his disposal. When using a medium, the user can operate only within the pathways laid down by the designer of the material being used. This path is obvious in the case of, say, a TV program, but even in the case of a sophisticated computer-based package, with multiple responses or choices possible, the student is effectively confined to a predetermined pattern. When using a *tool*, on the other hand, the restrictions on use are those inherent in the tool itself, and the user is free even to try to discover new uses for it. The benefits of teaching students to use the tool of computational analysis for themselves are closely related to this freedom; precisely because a tool contains no built-in promptings, the student must think hard about his approach to a problem and how he is going to use the tools at his disposal. This leads him to consider all the steps of problem-solving, but particularly interpretation as this is an essential part of the checking process in numerical work. By the time he has written his program and run it two or three times, he will usually have a good intuitive understanding of the properties of the mathematical model he is studying, and will be able to relate features of the solution to the various component parts of the model – essentially he will have interpreted his solution. The question of problem formulation also becomes more open, because features of the solution may suggest weaknesses in the original model or perhaps show that some feature originally included was not very important in practice.

The complex interaction between various parts of the process of solution is illustrated in the diagram in Figure 4.1.

Figure 4.1

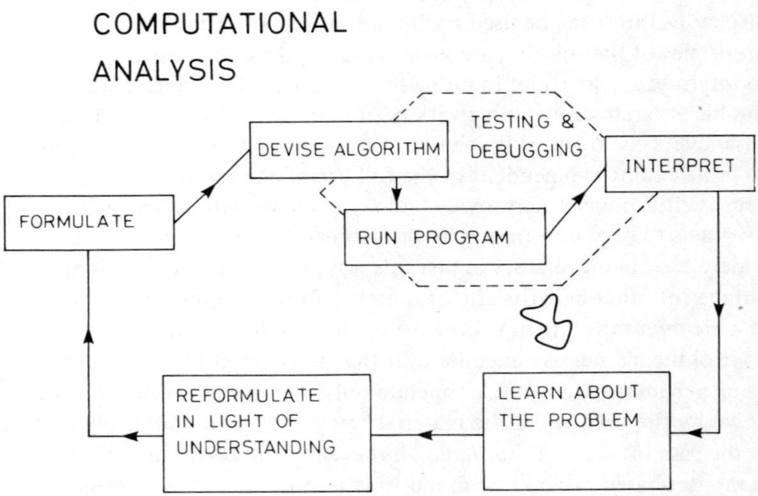

Some Disadvantages

There are disadvantages associated with computational analysis. The greatest is probably the fact that even with the most advanced programming languages, considerable attention has to be given to grammar and syntax, and so a substantial overhead of time is needed to put a problem on the computer. There is also the danger of a user becoming a 'computer freak' and failing to use other mathematical techniques. At this point it should be made absolutely clear that the

Figure 4.2

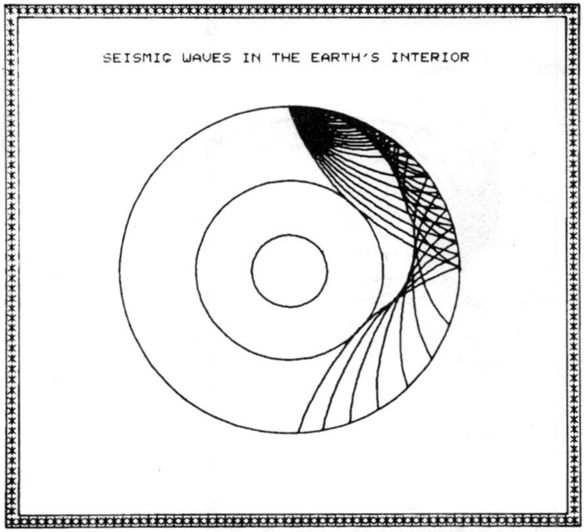
SEISMIC WAVES IN THE EARTH'S INTERIOR

(a)

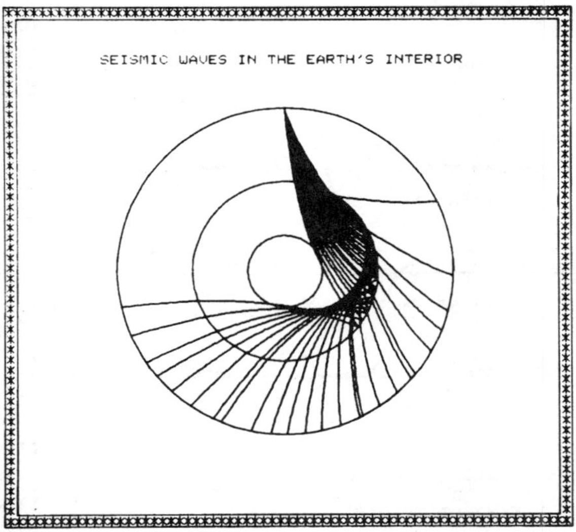
SEISMIC WAVES IN THE EARTH'S INTERIOR

(b)

Figure 4.2 (c)

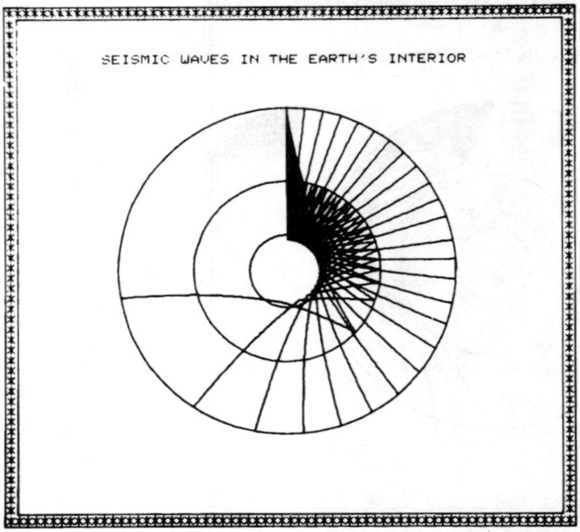

Figure 4.3

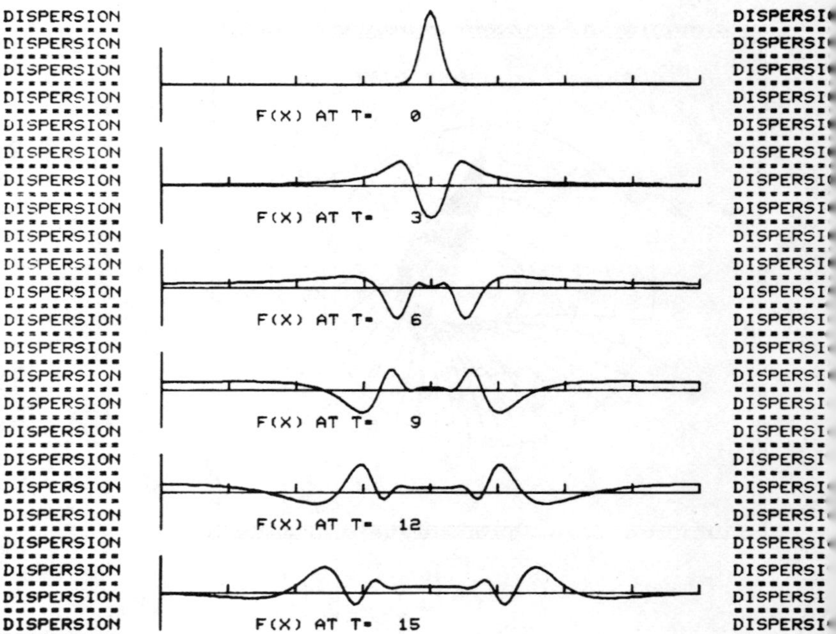

author does not favour *replacing* traditional techniques; the intention is to complement them by adding computational analysis to the mathematician's tool kit.

There are also problems in bringing computational analysis into the undergraduate curriculum, the most obvious being the scale and cost of suitable computing facilities. Other important problems are: which techniques to teach; how to teach them; and how to integrate them into existing course structure. A description of how this has been achieved at the author's institution has been described elsewhere.[1]

The technique can be employed effectively to encourage group working.[2] Briefly, a lecturer was giving an introductory course in Seismology. The central part of the course hinged on imparting a knowledge of how seismic waves propagate through the Earth. The lecturer organised his class into teams who constructed a variety of programs which simulated this. Towards the end of the course the teams demonstrated their program to the rest of the class and described what they had learnt. The illustrations (Figure 4.2) of some typical results show ray paths through the Earth.

Two further examples of the use of computational analysis will be described. They are both taken from amongst the set currently available to students at the author's department.[3,4]

The first example concerns the phenomenon of dispersion. Imagine the surface of, say, an ocean, on which a one-dimensional initially stationary sinusoidal disturbance is set up, for example, proportional to sin kx. Then it can be shown that this will propagate sinusoidally in time with frequency

$$\omega = \sqrt{gk}.$$

Now consider an arbitrary initial disturbance: this can be Fourier-analysed into sinusoidal components. As we know how each of these evolves in time, we can obtain a formal solution for the evolution of the initial disturbance. If f(x) is the initial disturbance, then the formal solution is

$$F(x, t) = \frac{1}{2\pi} \int_{-\infty}^{+\infty} \int_{-\infty}^{+\infty} f(s) \exp[iwt + ik(s-x)] \; dsdk$$

where $\omega = \sqrt{gk}\sin(k)$.

However, a much clearer picture is obtained from numerical solutions, as illustrated in Figure 4.3. Here the initial disturbance is a 'hump' and it will be seen that ripples spread out away from the initial position. By following individual wave-crests, it is possible to see that the group velocity of each packet of waves differs from the phase velocity. The program to produce these results is very simple, given the existence of a Fourier Transform library routine, and once written it can easily be adapted to other initial conditions or to show the effect of a different relationship between ω and k.

Our second illustrative example is the motion of a rigid body under zero forces, for example a satellite. This is an almost universal topic in courses on rigid body dynamics. All rigid bodies have a special axis system (the principle axes of inertia), and if the angular velocity of the satellite with respect to fixed axes instantaneously aligned with the principal axes is $(\omega_1, \omega_2, \omega_3)$, then the angular momentum will be $h = (A\omega_1, B\omega_2, C\omega_3)$, where A, B and C are the principal moments of inertia. The equations of motion of the satellite are known as Euler's equations and are:

$$A\dot{\omega}_1 = (B - C)\omega_2\omega_3$$
$$B\dot{\omega}_2 = (C - A)\omega_3\omega_1$$
$$C\dot{\omega}_3 = (A - B)\omega_1\omega_2$$

where dots denote differentiation with respect to time. The equations have two simple first integrals – energy and magnitude of h are constant – but the general solution is possible only in terms of elliptic integrals.

Figure 4.4 shows a typical numerical solution in graphical form. All three components of angular velocity are plotted against time, and in addition a perspective view of the motion of the point $(\omega_1, \omega_2, \omega_3)$ is given.

Evaluation and Conclusions

What evidence, it may be asked, can be found to justify the cost and effort that computational analysis entails? Evidence on this has been hard to acquire, although it should be remembered that educational evaluation of any degree course would be difficult. At his own instruction, the author has adopted two approaches.[5,6] The first relies on feed-back from students and staff (mainly through questionnaires), and the second relies on statistical analysis of examination marks. The broad conclusions to emerge are that there is positive appreciation by students for the use of computational analysis; there is a widespread feeling by staff that the idea is of great educational value, and that in terms of examination marks there is considerable benefit. The latter

Figure 4.4

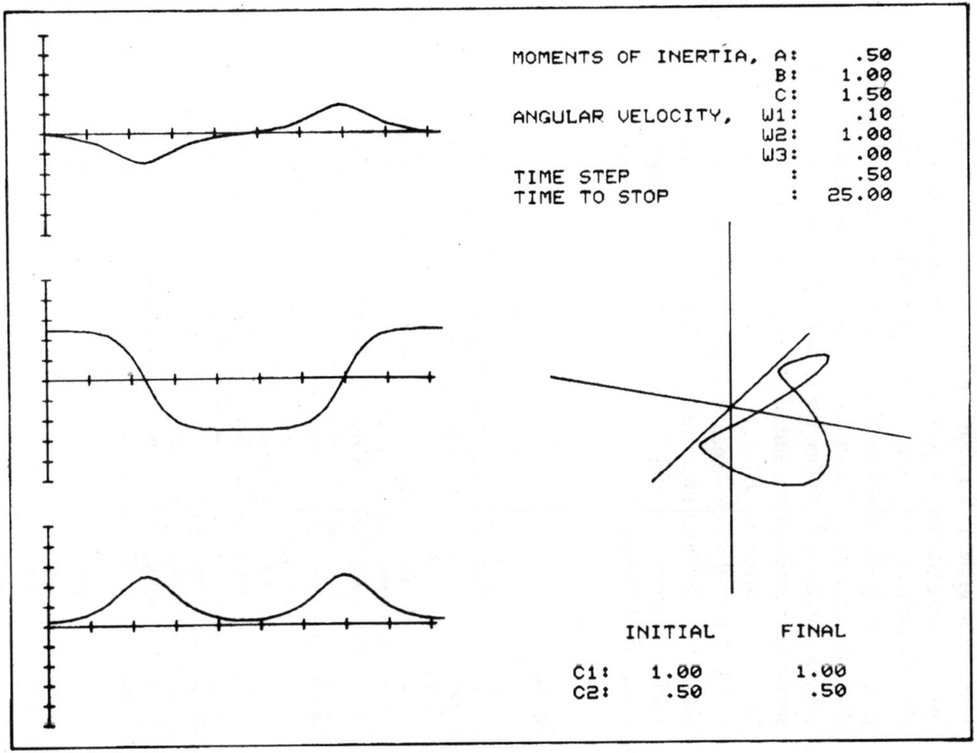

MOMENTS OF INERTIA, A: .50
B: 1.00
C: 1.50
ANGULAR VELOCITY, W1: .10
W2: 1.00
W3: .00
TIME STEP : .50
TIME TO STOP : 25.00
INITIAL FINAL
C1: 1.00 1.00
C2: .50 .50

point naturally exerts a strong influence on students' attitudes to the use of computational analysis and is particularly important in view of the relatively cumbersome software available for mathematical programming which means that computing work can absorb a great deal of time. Indeed it is in this area that the main disadvantage of computational analysis lies; the remedy must be sought in future software improvements, and some general thoughts on how this might be done are given in a later chapter.

In general, the experience gained to date is seen to justify fully the claim that computational analysis is a powerful new technique in applied mathematics, and that it ought to play an important role in university teaching.

References

1. R.D. Harding, 'Computer-aided Teaching and Applied Mathematics', *International Journal of Mathematics Education in Science and Technology*, vol.5 (1974) 447-55.

2. R.D. Harding, 'A Case Study of the Use of Computer Graphics as an Aid in Applied Mathematics Teaching', *Proceedings of the IFIP Second World Conference: Computers in Education* (Marseilles, September, 1975).

3. Computing Practical Class Manual for the Course 'Computational Projects in Applied Mathematics' (available from DAMTP, Cambridge).

4. *Applied Projects in Computational Mathematics*, Part 1 – Project Descriptions (2nd edn., November 1977, available from DAMTP, Cambridge).

5. R.D. Harding, 'Evaluative Development of a Computer Assisted Learning Project', *International Journal of Mathematics Education in Science and Technology*, vol.7, no.4 (1976) 475-83.

6. D.C. Johnson and R.D. Harding, 'The University Level Computing Experiences and Mathematics Problem Solving Ability'. To be published in *Journal for Research in Mathematics Education*, January 1979.

5 COMPUTERS IN TEACHING BIOMEDICAL SCIENCE

5.1 Introduction

Earlier chapters have indicated the important role played by computers and mathematical models in biomedical research and also how, in mathematics teaching, the computer can help in giving mathematical insight into physical systems. In the second chapter, some reasons for a lack of fluency in mathematical language amongst life scientists were discussed with some indications of ways forward. It is now appropriate to examine a few specific ways in which a new generation of scientists may be trained to express problems naturally in mathematical form when appropriate.

5.2 Teaching Biological Systems and Signal Analysis to Undergraduate Physiologists

R.I. Kitney

Over the last five years we have been running a course on the application of quantitative methods to the analysis of biological systems for final year undergraduates in physiology. The material for the course is mainly drawn from Control Theory and Signal Analysis. This immediately highlights one of the principal difficulties which arise in teaching, because in their original form both Control Theory and Signal Analysis, as taught in engineering departments, are highly mathematical subjects. Many attempts to teach quantitative methods to undergraduate life scientists have failed because the teachers have presented the students with what are essentially watered down versions of the corresponding engineering courses. In our experience a completely different approach is required. The two main aims of this new approach are: to maintain student interest at all times, even at the risk of covering less material; and to present material in such a way that the minimum amount is new. For example many statistics courses for life scientists use examples from physical science which are completely familiar to the physical scientist but which are totally new to the life scientist. The result is that the life science student is presented with two new sets of ideas rather than one, and worse, the example is pointless within the context of the course. Our experience has shown that in practice fairly difficult mathematical techniques can be mastered by physiology students with comparative ease provided that they can see the relevance of the technique to physiology.

Because of their background, it is assumed that the students attending the course will have little or no previous experience of quantitative methods. It is therefore essential to provide adequate grounding in the ideas and techniques which will be presented. To this end the course is divided into two overall sections:

1. The Qualitative Background

The general aim of this part of the course is to cover in qualitative terms the strategy which underlies the various methods which are commonly used in the study of biological systems from the systems and signal analysis point of view. The primary reason for this approach is that the vast majority of textbooks and research papers discuss at length how a particular procedure can be undertaken, but it is however rather rare for the author to explain why the method was chosen. In practice this is exactly the kind of information which is vital to a newcomer to the field. The physiology student is primarily much more interested in WHY rather than HOW. Hence having been convinced about why, there is usually good motivation towards discovering how the technique can be used.

2. Quantitative Analysis Techniques

Biological systems can be divided into three types for the purposes of quantitative analysis. These are: systems which exhibit a suitable signal for analysis, systems where an appropriate stimulus can be applied and an appropriate response measured, and regulatory systems. In the latter case the regulatory nature of the system often forces the researcher to resort to computer modelling.

The Analysis of Systems which Exhibit a Suitable Signal

The case of systems which exhibit a suitable signal for analysis are considered first because they are familiar to the student within the context of general physiology. All the students will have met continuous signals like the ECG or the respiratory wave-form, and point event signals like neural signals, although the terms continuous and point event will be unfamiliar. The next stage therefore is to categorise general biological systems and subsystems into four different types.

Systems with only continuous signals

Fully continuous signals: blood pressure wave-forms of various parts of the vascular system can be quantitatively related in order to describe features of the intervening vasculature; aspects of respiratory activity can be investigated to relate the pattern and depth of breathing to Po_2, Pco_2 and pH levels; the control of various hormones in the endocrine system can be investigated by the quantitative study of dynamic concentration patterns.

Systems with mixed point event and continuous signals

Mixed point event and continuous signal systems are met, for example, in the relation between motor-neurone spike-train potentials and the resulting skeletal muscle mechanical response (investigated for example in order to obtain the impulse response of the muscle *in vivo*); vasomotor and venomotor activity also comes into this category.

Systems with only point event signals

Fully point-event signal systems are, of course, illustrated by neuronal networks, often investigated primarily in order to elucidate functional neuro-anatomy.

Systems containing both continuous and point event signals

Mixed continuous and point-event signal systems are illustrated by biological receptors in general, particularly chemoreceptors, the evaluation of which is illustrated by the use of pseudo-random binary sequences stimuli and a cross-correlation procedure.

Having introduced the concept of categorisation of signals, the next stage is to introduce the idea of describing continuous wave-forms by mathematical equations. This is an important step because it confronts the student with two new concepts, the first being that wave-forms can be modelled by mathematical equations, and leading directly from this, that any continuous wave-form can be broken down into a series of sine and cosine waves of different frequencies. A very useful example is the respiratory wave-form which often approximates very closely to a simple sinewave. Knowing the amplitude (A) and the frequency (w), i.e. two parameters, the respiratory wave-form can be reproduced from the simple mathematical model Asin (wt). At this point it is also helpful to get the student to draw periodic wave-forms of different known frequencies, add them at different points in time to produce a complex wave-form, and check that the same complex wave-form can also be generated from the addition of the equations which represent the individual wave-forms.

The other important concept which must be introduced as early as possible is that of representing the signal in the Frequency Domain. Three parameters which will already have been introduced to describe signals are: frequency, amplitude and phase. These are normally associated with representing signals as a function of time, i.e. signal representation in the Time Domain. The alternative representation of the information contained in the signal, i.e. plotting it as a function of frequency, can be explained in terms of the three parameters already considered. Approached from this angle it should seem logical to the student that as there are three parameters, plotting the information contained in the signal as a function of frequency requires two graphs rather than one. Hence amplitude plotted against frequency becomes the amplitude spectrum and phase plotted against frequency the phase spectrum. Having introduced frequency spectra, it is worthwhile returning to why these techniques are useful in the analysis of the behaviour of biological systems. An appropriate way to make this point is simply to show examples of a biological signal plotted as a function of time under two different circumstances and simultaneously project the corresponding amplitude spectra. Providing suitable records are chosen, the information revealed by the spectra as opposed to the original wave-forms is often startling.

Practical signal analysis usually involves the use of digital computers. It is therefore necessary to discuss how the signals recorded during a biological experiment are converted into a form which can be handled by a digital computer. In other words how the signals are converted into sampled form. This procedure is usually referred to as analog-to-digital conversion (ADC) and an understanding of frequency spectra enables a much clearer understanding of the process. Experience has shown that when teaching the type of course we are discussing it is equally dangerous to include too little mathematics as opposed to too much because without sufficient mathematics the logical basis of methods is often totally lost. This is particularly true of analog-to-digital conversion because without a knowledge of spectra the procedure becomes extremely difficult to understand.

At this point in the course the student will have been confronted with a series of new ideas which follow from a consideration of how continuous wave-forms can be analysed. The other important area as far as systems which exhibit suitable spontaneous signals are concerned is point event signals, e.g. neural spike trains. The analysis of such signals is slightly more complicated than continuous signals, but hopefully the student will be able to build on the material already presented. A large number of analysis techniques associated with neural spike trains involve the interval between spikes, either time intervals between adjacent spikes or more complicated combinations of timings. In order to understand how the various analysis methods are carried out in practice one must have a knowledge of basic computer programming. The present course has been designed so that in its early stages computer programming is taught in parallel with the other material. Consequently by the time the analysis of neural signals is discussed the student will have a reasonable working knowledge of a high-level computer language like FORTRAN and understand how arrays are used in programming. It should therefore be relatively easy to explain how derived wave-forms like interval wave-forms are calculated from the original recorded signal.

Systems Analysis Using the Stimulus/Response Approach

In the categorisation of systems which we have adopted for teaching purposes, the second type of biological system which was defined is one where a stimulus/response experiment can be performed. Whether a complete system, or only part of it, is being analysed, the approach is one of describing that part of the system between the stimulus and the response as a black box. The rationale is that if a particular parameter of the input, i.e. the stimulus, is varied, then the appropriate change in the output can be observed, i.e. the response, and from this difference a measure of how the system has modified the parameter, determined.

If a particular system is being analysed from scratch, the explanation given to the student is that the first factor which must be considered is: what aspect of the input function does the system respond to? For example, does the system respond to the magnitude of the input or its frequency, etc. It is clearly fundamental to determine this information as it defines the particular analysis technique to be employed. (It is important to mention at this point that the system may well respond to several features of the input, e.g. both amplitude and phase.)

There is, however, a general strategy which is often employed. Having determined the parameter of the input signal to be studied, an input level for the parameter is set and the same parameter measured at the system output, usually under steady state conditions. The input level is then altered and the corresponding output parameter again measured. If this procedure is repeated for a range of values, the result is a table of values which represents the system response. In general, the next step is to plot a graph, or a series of graphs, and from them to determine a mathematical description of the system behaviour. In the case of computer modelling the method is taken a stage further and a computer program written on the basis of the mathematical equations. Although this is a very simplified form of strategy it nevertheless is the basis for a number of well-known analysis techniques. Commonly used inputs are step functions, ramp function, impulses and most important of all, periodic wave-forms.

Two of the most important analysis methods in biological systems analysis determine what is called the Frequency Response of the system. In both cases the method results in an identical mathematical description. The two methods are: Bode and Nyquist Analysis. Having taken the student through the basic strategy of such methods it is then possible to explain that in both methods the frequency of the input is varied (a sinewave) incrementally and the changes in output magnitude and phase measured under two constraints, i.e. that both the frequency and wave-shape of the output are identical to the input. The next step is to plot the information graphically and it is logical that as there are three variables, one independent and two dependent, that there should be two graphs representing the systems response; the first magnitude change against frequency and the second phase change against frequency. The final stage is to explain how the appropriate equations can be fitted to these graphs in order to define the frequency response of the system. A point worth noting is that even the two graphs used to calculate the frequency response will be familiar to the student as they are amplitude and phase spectra.

In a more mathematical course it would be convenient to continue this line of thought in order to discuss the link between Fourier and Laplace Transforms. In the present course, however, we introduce the Fourier Transform simply in terms of its application to biological signal analysis. This is usually done at the point where the analysis of systems by stimulus/response techniques has been introduced because it is important for the student to become skilful at moving from the Time Domain to the Frequency Domain and vice versa. This is readily achieved by considering the Fourier Transform in its 'Fast' form and treating it simply as a mathematical operation on the computer; which in practice it is, because it is called in the form of a self-contained subroutine. The student is taught the principle of Fourier transformation without the details of the algorithm. It is therefore possible to explain the calculation of Frequency Spectra in terms of manipulation of the real and imaginary terms generated by the Fast Fourier Transform (FFT) subroutine.

By explaining the use of the FFT in biological signal analysis, it is possible to return to one of the basic aims of the course, i.e. to maintain a strong link with physiology. In order to do this the students are given a series of practical analysis problems to carry out on real biological data which is usually generated by a human experiment undertaken during the course. This data will have subsequently formed the basis of a demonstration of analog-to-digital conversion and ultimately been punched out on to cards. Each student will therefore possess an identical set of data to his classmates and will understand where the data has come from. Examples of the type of problems which are set are:

1. The removal of drift using least squares fitting.
2. Cross-correlation and auto-correlation in the Time Domain.
3. The calculation of Amplitude, Power and Phase Spectra.
4. Fast auto- and cross-correlation, i.e. via the FFT (including a comparison between the results of correlation via the FFT as opposed to calculating the function in the Time Domain).
5. Cross Amplitude and Phase Spectra.

Systems Analysis by Computer Modelling

The third type of biological control system which we discuss involves those systems which perform a regulatory function. There are, of course, many examples of such systems in the body; hence the important factor in any discussion is to choose examples which are taken from systems which are familiar to the physiology student. Regulatory systems by their very nature are not conducive to stimulus/response techniques because the role of the system is to remove any perturbation. Consequently, it is often the case with such systems that the method of analysis is computer modelling. A good deal of the strategy involved will have already been covered in relation to stimulus/response methods. Nevertheless, the difference in strategy is sufficiently great to warrant a complete description. Once again, the initial point is made that the strategy which will be outlined only represents a local view of how to approach the problem.

The first stage is always to emphasise to the student that in developing a computer model of either a whole physiological system or part of it, the initial step is to gain a thorough knowledge of the relevant physiology. On the basis of this knowledge, it is possible to develop a block diagram of the system. To the physiologist a block diagram is probably simply a convenient way to show the relevant information. This is not the case when developing a computer model because the block diagram forms the plan on which the simulation is built. Hence, the various blocks and pathways which are shown represent the operating strategy of the biological system and these will eventually be converted into mathematical equations. Having discussed the structure of the block diagram, the students are encouraged to discuss how the characteristics of the various blocks might be determined. The aim of this discussion is to make it clear that many of the analysis techniques already described in the course can be used to determine the characteristics of the constituent parts of the regulatory system. Another possibility which is discussed is that individual blocks may well have been described by other workers.

Within the scope of the present course it is not possible to tackle complex simulations. The objective is therefore to discuss the general approach within the framework of the students physiological knowledge. Therefore, having outlined the mathematical description of individual blocks, the next two steps are: converting these descriptions into computer language, and linking the various parts of the resultant program together. In general, the language we use for this purpose if FORTRAN. While it is certainly not the most efficient computer language for simulation, the rationale is that it is certainly the most freely available language and, as such, will be of most use to the average physiologist. The final two stages involve the iterative process of running the model and comparing the results with known data, and modifying the model in the light of this comparison.

Concluding Remarks

In summary then, the aim of the course is to give final-year physiology students a perspective on the use of systems and signal analysis techniques in the study of biological systems. The approach is always one of introducing new material by way of examples which wherever possible are taken from theory which is familiar to the student. A general strategy for the analysis of biological system is provided and at each stage emphasis is laid on WHY a particular technique is used. This basic strategy falls into three categories: methods which can be used to analyse existing signals, stimulus/response techniques, and computer modelling.

A great deal of modern research involves the use of the multi-disciplinary approach and hence, by definition, it is likely to involve the physiologist in working with physical scientists, mathematicians and computer specialists. Our objective is therefore to enable the physiologist to work successfully with other scientists and to ask searching, relevant questions.

5.3 Transparent 'Black Boxes'

A.W. Wood

Mathematical models of biological systems or processes, when presented in the form of a computer simulation, can form the basis of an effective way of teaching those systems or processes. A student can see the effect of altering the values of variables in the model in a way that is not always easy in the real-life system. In particular, one variable can be altered independently of the others to demonstrate effects which would otherwise have to be inferred from a large series of 'real' experiments. The use of such simulations is of course very helpful if the 'real' experiment is unavailable because of such factors as difficulty of technique, expense of reagents, the undesirability of animal experimentation or the length of time that would be required to achieve the experimental goal. They need not be used as a replacement for experiments: computer simulations can be used as an adjunct or as a preliminary to doing experiments on the 'real' system, to allow the students to develop an investigative strategy quickly and effectively. This of course leads on to the use of models as part of the scientific method of creating, testing and modifying hypotheses. The student can, by using the computer, compare the behaviour of more than one model and evaluate them for 'goodness of fit'.

The whole process of interpreting output from the computer simulation is made much more efficient in these kinds of activity by having a pictorial or graphical display rather than a table of numbers. In general biology and medical students tend to favour qualitative rather than quantitative descriptions, and their level of mathematical expertise and enthusiasm for numbers is not high. The question is: can the use of such simulations, with graphical output, stimulate non-numerate students to take an interest in mathematical modelling, now that the mathematical concepts can be visualised pictorially, and the drudgery of computation can be removed? The educational context in which these simulations tend to be used certainly favours this. The most common use of simulations educationally is in the form of computer-assisted learning packages consisting of one or more programs covering a particular topic and some associated written material usually in the form of a student guide. As much information as is deemed necessary concerning the model can be included in the written material, including listings of parts of the program. Usually, students work through the packages in pairs or small groups, interpreting and discussing the output with each other in conjunction with the written material. Conversations and activities of students at the computer terminals have been analysed by Laurillard.[1] This study showed that 'learning promoting' activities such as checking knowledge and understanding, interpreting, experimenting and reasoning were encouraged, which led to gains such as visualising and understanding.

How can this favourable environment be channelled in such a way as to focus the attention of life science students on to mathematical descriptions? Most student guides to biology packages are written in such a way that the model remains concealed from the students: he (or she) uses the program as a 'black box' to learn about a biological system. This does not mean that the student gains no picture or 'model' of the system; he will, as the simulation is explored, build a qualitative 'thought model' of how the system behaves, and will 'get a feel' for that. What is desirable is that these students, who are after all being trained to be scientists, should learn to be able to put numbers to their 'feeling for' the system, and be able to describe some of the features in mathematical terms, as indeed the underlying program does. The problem, then, is how can these black boxes be made transparent to the user – how can 'windows' be opened up – without at the same time alienating the user because of his or her lack of mathematical expertise?

Let us now look at some specific CAL (Computer Assisted Learning) packages to see how an answer to this might be possible. These are physiology packages from the Computers in the Undergraduate Science Curriculum project.[2] The first of these aims at teaching the indicator dilution technique for measuring cardiac output[3] (a technique which may cause animal suffering if performed by inexperienced hands). The principle behind this technique is also rather hard for students to grasp and difficult to teach by conventional means. The simplest model assumes that the heart acts like a reservoir with constant inlet and outflow, and that dye substance can be added instantaneously to blood in the reservoir, and that these then rapidly and uniformly mix. The concentration of dye in samples withdrawn from some point in the outflow tube (an artery) will look something like Figure 5.1: a discontinuous rise followed by an exponential fall-off.

In fact, dispersion of dye, the time taken to inject it, recirculation

Figure 5.1 Dye concentration curve for idealised system. A: amount of dye added (mg); V: volume of reservoir (l); F: flow rate in outflow (or inlet) tube (l/sec).

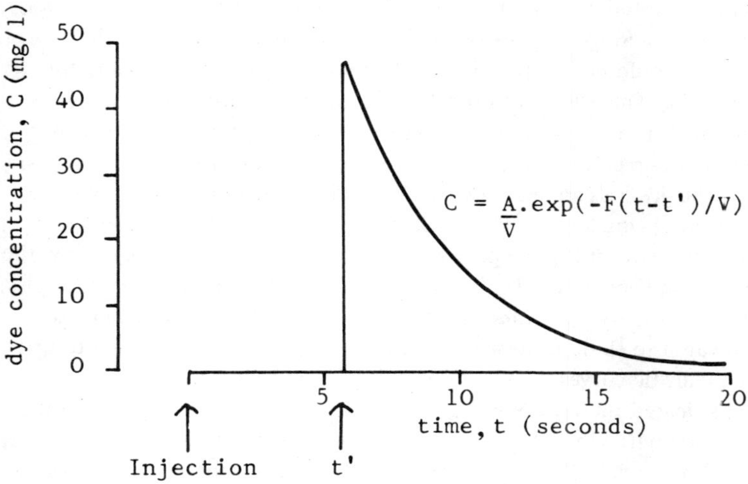

Figure 5.2 Typical dye dilution curve (dog heart)

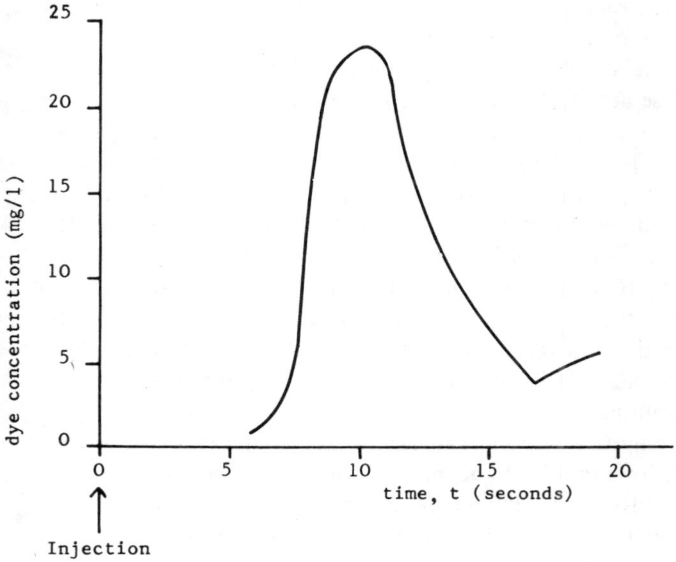

and random factors make actual dye dilution curves look more like Figure 5.2. This illustrates the problem well. In a 'black box' simulation, curves like that in Figure 5.2 could be output by the computer by having no model at all in the program. Values from actual experiments could be stored in computer files and retrieved under program control. The user could use any curve to calculate cardiac output, and this estimate could be compared with the previously calculated value for that curve. The effects of doubling the amount of dye added, or the rate of addition could also be demonstrated by the program retrieving different curves and matching the desired changes. A simple theory could be given as an appendix to the student guide. But 'windows' are not really being let into the black box, if by black box one means the program. After using the package the student is still unaware of how the curves were generated. He may indeed believe that a modified form of the given theory was embodied in the program. If he was sufficiently motivated to find out how the simple theory could be modified to give such realistic curves, he would be disappointed. Some quite sophisticated theory does of course exist,[4] but it would be outside the undergraduate physiology curriculum to study this in detail. There would perhaps also be a problem with the size of program or the speed of computation if the most complex models were used. There would probably be a problem in size of program with the model-less type of simulation, because a curve approximating to each combination of values of the variable factors would have to be stored by the computer, although interpolation would be possible.

The model chosen as a basis for the actual package represents a compromise between a model which allows a qualitative understanding of the system, and a model which provides a realistic representation of the system. The package in fact consists of three programs, the first being the simple model illustrated in Figure 5.1. This acts as an introduction, and the user can use this kind of display to study the characteristics of washout from a simple compartment – he can measure the time constant, change the amount of added dye, convince himself that the area under the graph is equal to the amount added divided by the flow rate. The notes invite the user to consider why the curve is the shape it is and to investigate various properties of exponential functions. The second program has a model which simulates empirically the initial dispersion and later recirculation of dye, and allows the dye to be injected over a finite period. A simple exponential is used for the major part of the rising portion of the curves as well as for the falling portion. The user can alter only the amount of dye injected and the frequency of taking sample, but his main task is to correct for the recirculation component by learning to extrapolate the exponential down-slope by taking logarithms Figure 5.3). The third program allows such factors as the heart size, volume of circulation, injection rate, position of injection and sampling and the cardiac output itself to be altered. The curves are superimposed on the graphics display for immediate comparison. The changes in shape of these curves can be related to the changes in parameter values in the model and to what would be expected in real life. The user can then begin to visualise the system in mathematical terms and appreciate the strengths and weaknesses of the model. He can look at the assumptions on which the model is based and appraise them.

Figure 5.3 Display of dye dilution curve from computer-assisted learning package. The terminal (Tektronix 4010) has a moveable crosswire which allows the co-ordinates of the points to be printed in the margin.

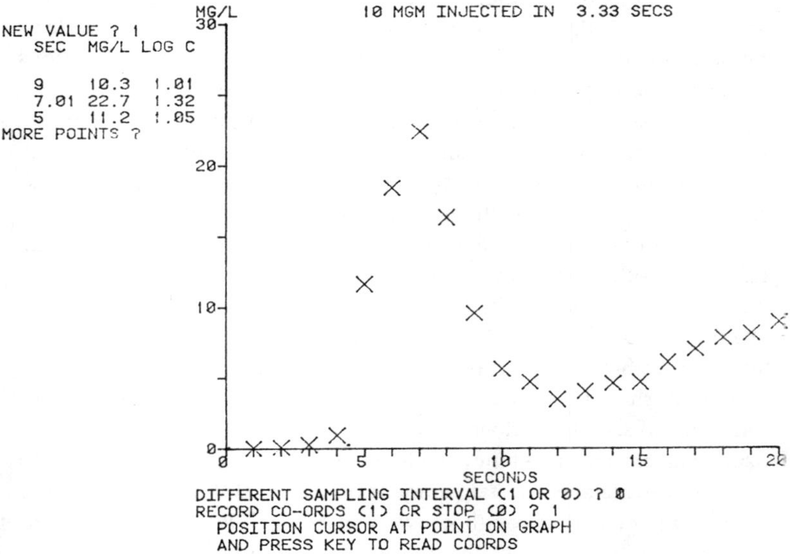

In the case of another package which teaches the autoregulatory properties of the isolated heart (the Starling heart-lung preparation),[5] the model used is not of primary importance in appreciating the principle of the way the heart muscle works. The program in fact is an assembly of simple 'Ohm's law' type relationships between pressure differences and flow, and empirical relationships using results from original papers.[6,7] The main use of this package is to act as a replacement for an 'unavailable' experiment, and to enable students to discover strategies for investigating the effects of a large number of variables, some dependent, some independent. However, the main equation in the model is a cubic in F, the blood flow, which is solved in the program by Newton's approximation method. The derivation of this equation is straightforward; a single starting value gives a wide range of correct solutions and the iterations converge rapidly. While it can never be claimed that mathematical methods become easier for biologists to understand by sticking biological labels on the variables, a graphical explanation in the student notes of what the program is doing and, perhaps by modifying the program to print a value at each iteration, the method of numerical solution should become plausible. Numerical methods are of course very common in research applications of the computer to biology.

A third package on counter-current mechanisms in biology[8] has as its basic model the heat exchanger equations well known to engineers.[9] This illustrates the use of analogues in biology – inferring mechanisms in ill-defined biological systems by comparing them with the properties of well-defined analogous systems as discussed in Chapter 2. The effects of altering flow rates or heat transfer rate between the tubes are investigated by interacting with displays similar to Figure 5.4. The improved efficiency of a counter-current compared to a parallel flow arrangement can be estimated. In the package, a second model is adapted to show how the counter-current mechanism can simulate the behaviour of specific biological systems, the whale flipper, the fish swim-bladder, and the kidney nephron. Assumptions and simplifications have to be made and a certain amount of empirical 'tinkering' is needed to make the model tally with known behaviour.[10] Would examination of these particular empirical approaches teach the user anything about mechanism? Perhaps not, but on the other hand in making assumptions and simplifications certain putative mechanisms have been in the originator's mind. This is all part of the scientific method, and it is perhaps worthwhile to 'window' into that.

The final package to be considered teaches about the transmission

Figure 5.4 Display from package on counter-current mechanisms. The student can change any or all of the listed factors by typing in a code letter and then a new value. The fluid travels from left to right in the tube represented by the upper curve, and in the opposite direction in the other tube (crosses).

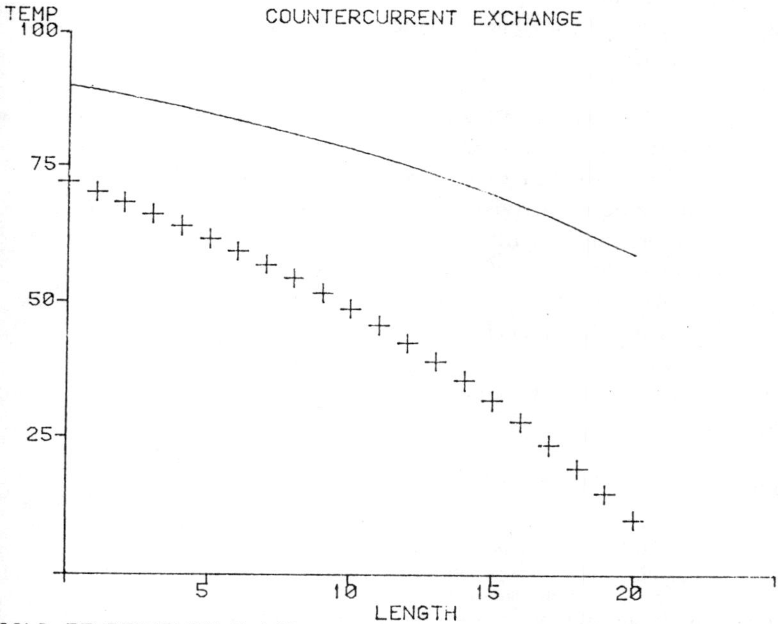


```
COLD TEMPERATURE ? 10C
HOT TEMPERATURE ? 90C
TRANSFER FACTOR ? .05
FLOW RATIO ? 2
LENGTH ? 20
 ?
```

of nervous impulses.[11] It both acts as an introduction to a sophisticated technique: voltage clamp studies of the nerve membrane, and as a way to visualise the theoretical description of the process: the Hodgkin-Huxley equations.[12] This package was designed at the outset to have large windows into the model since the model itself and its interpretation on the molecular level, is described in most textbooks. There is in fact a 1:1 correspondence of equations given in the student notes to the equations in the program. The display helps students to visualise the components of this fairly complex theory by encouraging them to make rough estimates of certain parameters from one display which can then be compared with another display, in which these parameters are computed. The interpretation of the shape of the curves is an essential part of the theory, especially in the linking of these curves to certain mathematical expressions arising from hypothetical molecular mechanisms. The biology student by interacting with graphical displays of these mathematical expressions learns to appreciate their significance.

In conclusion it can be said that it is instructive for biology students to be aware of the nature of the mathematical models used in biology CAL packages. In addition to the facility of being able to visualise the behaviour of mathematical expressions and relating this to the 'real' experimental conditions, looking at the methods of model building, with its assumptions, simplifications, empirical modifications and eventual integration is also of great value to the student. Further, models can also be critically evaluated for usefulness and 'goodness of fit' by supplying 'real' data. The educational potential of CAL is therefore greatly increased by having areas of transparency or 'windows' in black-box simulation programs.

References

1. D. Laurillard, *CUSC Evaluation Report* (Institute of Educational Technology, University of Surrey, Guildford, 1977).

2. Computers in the Undergraduate Science Curriculum (CUSC) project. A collaborative interdisciplinary project, financed 1974-77 by the Council for Educational Technology UK at University College London, Chelsea College London and the University of Surrey. Materials now available from the Educational Computing Section at Chelsea College.

3. 'Cardiac Output by Indicator Dilution' – a computer-assisted learning package from the CUSC project (Chelsea College, London, UK, 1978).

4. C.W. Sheppard, *Basic Principles of the Tracer Method* (J. Wiley, New York, 1962).

5. 'The Circulation: Isolated Heart-Lung Preparation' – a computer-assisted learning package from the CUSC project (Chelsea College, London, UK, 1978).

6. D.S. Riggs, *Control Theory and Physiological Feedback Mechanisms* (Williams & Wilkins, Baltimore, 1970), pp.419-29.

7. S.W. Patterson and E.H. Starling, 'On the Mechanical Factors which Determine the Output of the Ventricles', *Journal of Physiology*, 48 (1914) 357-79.

8. 'Countercurrent Mechanisms' – a computer-assisted learning package from the CUSC project (Chelsea College, London, UK, 1978).

9. A. Schack, *Industrial Heat Transfer* (Chapman and Hall, London, 1965).

10. K. Thurau and P. Deetjen, ' Die Diurese die Arteriellen Drucksteigerungen',

Pflügers Arch. ges. Physiol., 274 (1962) 567-80.
11. 'Voltage Clamping Technique' – a computer-assisted learning package from the CUSC project (Chelsea College, London, UK, 1978).
12. A.L. Hodgkin, 'The Croonian Lecture: Ionic Movements and Electrical Activity in Giant Nerve Fibres', *Proceedings of the Royal Society* (London), ser B, 148 (1958) 1-37.

5.4 Confronting Students with Mathematics

Two methods of involving biomedical students with mathematical ideas and techniques have been discussed in earlier parts of this chapter and the importance of ensuring immediate relevance was stressed in section 2.1 (p.15), Chapter 2. In the intervening papers, the key influence of mathematics and computing on research in the life sciences and methods of educating mathematicians have been considered. To what extent can the approach to using the computer as a tool in mathematics education be used in teaching students of the life sciences to become fluent in the branches of mathematics that are relevant to their study and understanding of science?

The bridging of the gap between the use of black-box simulation in computer-assisted learning and the computational tool of the mathematician would appear crucial. If any small steps can be made in the direction of science students building their own models by providing them with tools for 'experimental mathematics', then these may be crucial in the wider appreciation of quantitative methods by scientists with the accompanying deeper understanding of their science.

6 MAN-COMPUTER COMMUNICATION

6.1 Introduction

The need for a fluency in mathematics has been stressed and it has also been suggested that computer systems may be able to assist in this both at the level of education and research. Further, the use of on-line computers at both levels, which are in fact regions of a continuum, has been shown to be central to the development of science itself.

Both these points demand a means of communication between computer and scientist/student which is natural for the latter. Most present-day computer languages would appear to be inappropriate and the development of computer software has certainly lagged behind that of hardware.

6.2 Towards a New Computer Language

R.D. Harding

Inadequacies of Standard High-level Languages

The shortcomings of existing languages are not usually felt by the student who is using the computer as a medium for learning, in the sense described in the final section of Chapter 4. The reason is that such a student has been restricted in his use to the pathways laid down by the author of the package, and if this has been well written, every type of use will have been foreseen. If, however, the student wishes to begin using the computer more as a tool, then he is likely to be hindered by the nature of the computing language used to write the package in the first place.

Such a change of use is quite possible, because despite the distinction drawn in Chapter 4 between the computer as a medium and the computer as a tool, there is really no hard and fast dividing line. A student with an enquiring mind, after using a package, may well want to adapt it to solve a slightly different problem from that originally considered, and if the philosophy of Chapter 4 is accepted, he should be encouraged to do just that. The ideas of Chapter 5 and the concept of increased transparency are closely related too. The difficulty will usually be that to adapt a package will require a whole order of magnitude increase in computing knowledge.

If one begins by training the student to use the computer as a tool, barriers of this kind will not be so abrupt. However, there may not be time to train students in this way, and perhaps not all students will be capable of achieving the required degree of proficiency. Even for the expert, putting together a program in FORTRAN or BASIC is not a completely trivial operation; to achieve 'scratchpad' computing requires almost that: it should be possible to sit down at a terminal with a few rough equations and in five or ten minutes get the computer to produce some sensible and illuminating output. The benefits of being able to do this are exactly the benefits claimed earlier for computational analysis: essentially the computer should take the drudgery out of the solution step of problem-solving, enabling greater attention to be given to formulation and interpretation.

The ability to carry out 'scratchpad' computations is the next advance that should be sought in the progress of software. It follows naturally from the development of applications of computing since their invention.

Historical Perspective

In what could be called the First Age of the computer, computers were used as a last resort for handling tasks that could be done in no other way (in a reasonable time). Obvious examples are the solution of difficult differential equations, engineering design problems, optimisation of industrial processes, etc. In the First Age, computation provided the hard numerical answers that were called for. The first step in this process was always the construction of an algorithm, based on well-understood principles. The algorithm and the numerical results were not thought to add anything to the original understanding of the problem.

In the Second Age, where we are today, we have advanced to the point in which we compute not just for numerical results, but also for the insight that they can give into the original problem. Virtually all CAL projects in the United Kingdom (and no doubt elsewhere) make claims of this nature. Obviously First Age computing remains important, and continues to take up the bulk of computing activity, but a new element is present. As further evidence for this trend, the growing demand for graphical display devices can be cited.

However, Second Age computing remains restricted because of two barriers. The first is obviously cost, and the consequently limited resources that cost considerations bring; it is a widely accepted view that the future will bring improvements in this field. The second barrier, as already seen, is software.

The Third Age of Computing

It has been argued that there is a need for a language that is more suited to setting up models or solving problems; principally in mind are subjects where mathematical formulation is usual, but other subjects should not be excluded.

Progress has in fact been made in this direction: the software developed for the MATLAB project is a very good example.[1] Statements allowed include the following examples:

```
GRA X ↑ 2   FOR X = –5(1)5
```

plots a graph of X^2 for a range of values of X.

```
INT 4*T ↑ 2 – T   WRT T FROM 0 TO 1
```

carries out a definite integral.

MATLAB was designed to assist in the service teaching of mathematics at Further Education level, and not for setting up problem-solving packages, modelling or scratchpad computing. On the other hand in scientific computing today there exist an extensive range of powerful library routines, which will need to be 'glued' together with user written code. To do this the user has to be familiar with the appropriate language and the exact order and significance of the arguments needed by the routines.

What is needed is a universal syntax on the lines of MATLAB but encompassing almost any library routine available. Ideally, the calling sequence for each routine should be self-documenting, so that if a variable needed had not been set in advance, the system would interactively ask the user to supply a value, or perhaps an algebraic definition.

For example, the following should be acceptable:

```
SOLVE d²Y/dt² + k dY/dt + w² Y = F(t)
FOR Y(0) = 1, dY(0)/dt = 0, t = 0 to 2
AND PLOT Y AGAINST T
```

If k and w had not been specified, the system would respond with:

VALUES FOR k, w?

and if F(t) had not been previously defined, the system would ask:

DEFINITION OF F(t) NEEDED:

to which the user could reply: F(t) = a sin wt.

Besides requiring values of variables relevant to the problem, modern library routines often demand a number of control values or parameters: for instance, how much precision should the routine work do? An array of default values should be available in which the system could look up any values not supplied by the user: if it failed to find one that was needed it would then ask the user to supply a value.

A Multi-level Language

It would be quite feasible to write a system along the above lines – and indeed MATLAB is a step in this direction – but it has still not gone far enough to meet the difficulty discussed above, that there is an abrupt barrier to be crossed as soon as it is desired to alter the capabilities of the system in some way. To deal with this problem it would be necessary to devise a language that was universal to the whole system. At the highest level it would be the problem-orientated package system we have discussed, and users would enter the system at this level by default. The next level down would contain the software that was driving the highest level. If this used the same syntax and systems for defining data as the highest level (but of course using a completely different set of routines), then it would not be such a formidable barrier for users to cross if they wished to alter the facilities available at the highest level in some way. In principle a whole hierarchy of levels can be envisaged containing routines which would deal with every aspect of the whole computer system, from I/O handling resource allocation, etc. up to the user level that has been discussed.

The greatest advantage of such a language structure would be that the language itself would cease to be static. Each user would be more inclined to adapt the language a little to suit his own needs: improvements that were useful to others could easily be made generally available, and an evolutionary phase of language development would set in.

Conclusions

In principle the idea of a multi-level language would, if implemented, be a major step forward. It is also clear that there are considerable problems in specifying the idea closely enough for it to become a reality. The main conclusion to be drawn is that whilst one might agree on the shortcoming of present-day software, it is difficult to see a way to making a start on devising a suitable syntax and data structure for the putative multi-level language. It seems clear that it would be necessary to look for progress at a less ambitious level, the obvious line of development being an extension of the MATLAB concept to cope with a wider class of applications, and to adapt it to be more useful as a language for writing simulations and to deal with default values in standard library routines. It is hoped, however, that the more ambitious objective of a multi-level language should be kept in sight and perhaps one day achieved.

Reference

1. D.F. Leach (1974), 'A computer-based laboratory for service mathematics teaching'. Proceedings of the 1974 International Conference on Frontiers in Education, London, IEE ISBN 0 85296125 1.

6.3 The Delay Concept in Computer Simulation Languages and the Role of These Languages in Modelling and Education

F. Oliveira-Pinto

Introduction

Much has been written on the applications of computer simulation languages to the description of complex abstract models of certain aspects of reality but very little has been said about their contribution to the development of the basic concepts of abstract modelling. (Kiviat's paper[1] is one of the few exceptions on this point.)

Modelling, like creative writing and theorem proving, is a highly demanding mental process and one of the possible ways to look at its internal structure is to analyse the abstract concept which computer simulation languages have established through extensive experimentation.

If a particular modelling concept has become popular in simulation practice it is almost certainly useful. And although the concept may not be essential in model construction it will certainly deserve consideration because it may lead to a more fundamental or powerful tool for system analysis.

In the following we shall look into basic concepts underlying some of the more widely-used computer simulation languages in particular to see how they can provide a framework for creative model description.

Computer Environments

Computer environments offer the resources required for the implementation and testing of modelling techniques because of their operating flexibility and ease of problem standardisation.

Historically, general purpose computer languages were used first to describe complicated modelling situations but soon it became apparent that they suffer from the same shortcomings as natural human languages: lack of precise and concise statements to describe essentially dynamic situations. Emshoff and Sisson[2] report that for the modelling of the scheduling mechanism of a new computer operating system still in construction and using FORTRAN IV (which they hoped to optimise prior to its release) it took four man-weeks to program and three man-weeks to correct: the FORTRAN description was about 600 lines long.

They repeated the same exercise with very minor modifications using the computer simulation language GPSS (GPSS stands either for General Purpose Systems Simulator or more recently for General Purpose Simulation System) and found that it needed only one man-week to program and check and the GPSS description was only 30 lines long.

The reasons behind such widely different performances are due to the fact that high-level, problem-oriented general purpose languages transfer the control of computer operations either sequentially or by direct logic transfer, while simulation experiments rely instead on the ordering in time of the relevant system events. In other words, the built-in logical control routine of the general purpose language is not able on its own to control the time flow of system events and a *master clock* or *time control routine* needs to be provided. The difficulty is that the construction and optimisation of such time control routine is not a trivial job and therefore in most cases will be time-consuming. Further, it is also very important the way the simulating system gains access to the system control lists either to delete or insert new items of information. Since general purpose languages do not usually provide such flexible data structures they are not immediately suitable for such information retrieval operations.

Specialised job simulators for management and military applications, which share the advantages of human-like language description, have also been developed, but their lack of flexibility for problems outside their limited design range makes them unsuitable for general theoretical considerations.

The first interpretive general purpose language which was developed in England around 1960 by Kelley and Buxton[3] was limited to MONTE CARLO experiments and named MONTECODE. It had statistical facilities and random number generators but limited scope for event serialisation. Others followed immediately with different conceptualisation to system description and their implementation in computer environments. Before we embark on their study let us first characterise their problem domain.

Discrete Systems

We shall limit our problem domain to the study of systems where we can define a one-to-one mapping of system changes into the values of one of the system variables. For convenience we shall associate in this description this reference variable with the system time component and we further assume that the sequence of time instants is denumerable. Thus we shall consider the description of DISCRETE SYSTEMS, namely systems whose behaviour in the time domain can be completely described by a sequence of distinct system states. A system state is characterised by the numerical values of all its parameters and a system EVENT is the cause behind the change of a system state into another system state.

The concept of similarity between the simulated system and its model is then greatly simplified because it rests entirely on the 'distinguishability' of particular situations or states in both systems and on their correspondence. We shall then look at alternative but equivalent ways of describing system states as time evolves.

In DISCRETE systems the concept of EVENT is therefore fundamental. We may associate sequences of EVENTS, ordered in time, as representing specific system PROCESSES: or we may take a collection of elementary propositions (logical or otherwise) leading to, or including, a change in a system state and call it an ACTIVITY.

Both of the above alternatives have advantages in particular applications. Depending on the stress we put on them we have:

1. EVENT SCHEDULING approach which emphasises the EVENT as the driving force in system changes;
2. PROCESS INTERACTION approach which stresses the existence

of distinct entities within the simulated system;

3. ACTIVITY SCANNING approach which reviews collections of elementary propositions (logical or otherwise) which can begin or terminate each time an EVENT takes place.

The development of general purpose computer simulation languages has followed these three concepts (or some combination of them) and among the most widely used we mention: SIMSCRIPT: originated by Markowitz, Karr and Hausner[4] using the most elementary EVENT SCHEDULING approach; GPSS and SIMULA originated respectively by Gordon[5] and Dahl and Nygaard[6] using the more ambitious PROCESS INTERACTION approach; and, finally, Control and Simulation Language (or CSL for short): originated by Buxton and Laski[7] using the ACTIVITY SCANNING approach. All these computer simulation languages have undergone important modifications and extensions thereby evolving substantially from their original design specifications. We shall take these major alterations into account in the brief characterisations which follow.

Simulation Languages

The GPSS Language

GPSS is the oldest of computer simulation languages still in general use. It assumes limited previous computer programming experience or none. As a computer language it is primitive in design and has limited syntax flexibility. Nevertheless it possesses powerful statements which enable the inexperienced user to describe rather elaborate simulation situations in a simple and laconic manner.

To allow explicit interaction amongst different system PROCESSES, GPSS introduces two distinct concepts of time delay: *conditional delays* and *unconditional delays*.

1. *Conditional delays* are of undeterminable time duration because their termination is entirely dependent of the relaxation of temporal system constraints. Since system constraints can only be changed by a different system state it is the competence of the GPSS scheduling routine to check automatically whether or not conditionally delayed PROCESSES may proceed whenever a new system EVENT takes place.

2. An *unconditional delay* is, on the other hand, a straightforward time delay whose duration is known in advance. It acknowledges the existence of fundamental and unavoidable system activities.

The SIMULA Language

SIMULA is a general purpose programming language developed around the well-established ALGOL 60 algorithmic language (but unfortunately not having it as a subset). SIMULA 67 (an improved version of SIMULA) takes the ALGOL 'block' concept* a step further and defines super-blocks called *classes.*

With the *class* concept program control may re-enter the ALGOL 60 blocks anywhere within the block (provided that some elementary precautions are taken) and therefore interference from outside classes and blocks is possible and legitimate. If, in addition, these classes are prefixed by the work PROCESS then they exist under the supervision of the SIMULA monitor and, if they describe simulation PROCESSES, they can act as detached components or simulation OBJECTS.

From the simulation point of view SIMULA replaces the GPSS conditional time delays by *undefined* delays but keeps the *unconditional* GPSS ones. Undefined time delays are basically different from the conditional ones because their termination is obtained by direct user action not by system automatic retrieval. Thus the scheduling mechanism is simplified but the model description becomes more complicated.

The CSL Language

CSL on the other hand eliminates unconditional time delays at the expense of an even more complex system description. Without unconditional time delays there is no EVENT scheduling activity and the time differences are replaced by logical propositions more in line with other general purpose programming languages. CSL time delays are therefore always conditional and the language control routine scans through all available ACTIVITIES whenever the current ACTIVITY terminates.

The SIMSCRIPT Language

SIMSCRIPT uses the pure EVENT approach and concentrates on scheduling. That is, in fact, logically equivalent to the unconditional time delay concept referred to previously.

*By a 'block' language is understood a language where not only closed and environment-free subroutines can be implemented but also where closed regions within the language can be implemented and administered as if the presence of the surrounding text did not matter.

Computer-aided Teaching

A wide range of simulation models is used in teaching projects and it is customary to require students to interact with predefined models either using the different numerical values for the model parameters or selecting eye-catching graphical forms for the output of their results. The next logical step in computer-aided teaching projects is to provide the more resourceful student with the possibility of modifying or extending existing teaching models. This ultimate aim may sound unnecessary, or too pretentious, but it is well known among active computer users that the more novel and ingenious contributions frequently come directly from the students. It is therefore important that the language in which simulation models are described is high-level and problem-oriented with concise and powerful statements that simplify the problems of comprehension for the novice.

On the implementation side these simulation languages should be modular in structure (to allow the student simply to add or subtract something to an existing model) and interpretive in character (to allow the student to inspect and understand the source language whenever required). There is also considerable advantage in having the multi-access computer operating system written in its own simulation language since the users' programs are in fact interacting PROCESSES in the real world of the computer resources. If this were the case we should have reached the stage where the simulated system contained its own simulating model.

References

1. P.J. Kiviat, 'Development of Discrete Digital Simulation Languages', *Simulation*, 8 (1967) 65-70.

2. J.R. Emshoff and R.L. Sisson, *Design and Use of Computer Simulation Models* (Macmillan, New York, 1970).

3. D.H. Kelley and J.N. Buxton, 'MONTECODE, An Interpretive Program for Monte Carlo Simulations', *Computer Journal*, 5 (1962) 88-93.

4. H.M. Markowitz *et al.*, *SIMSCRIPT: A Simulation Programming Language* (Prentice-Hall, New Jersey, 1963).

5. G. Gordon, 'A General Purpose System Simulator', *IBM Journal*, 1 (1962) 18-32.

6. O. Dahl and K. Nygaard, 'SIMULA: An ALGOL-based Simulation Language', *Communications of the Association for Computing Machinery*, 9 (1966) 671-8.

7. J.N. Buxton and J.G. Laski, 'Control and Simulation Language', *Computer Journal*, 5 (1962) 194-9.

6.4 An Approach to the Design of Problem-oriented Languages, with an Application to Neurophysiological Experiments

N. Parezanović, B. Janković, Z. Damjanović

Introduction

Training people from different professional backgrounds in the use of computers at the level of algorithmic languages, such as FORTRAN and COBOL, meets with considerable difficulties. These difficulties arise primarily from the algorithmic nature of reasoning required by such languages which is not easily acceptable by non-professionals. On the other hand, problem-oriented languages, as is well known, are usually restricted to a specific field of application.

This section presents an approach to the design of language, which combines some advantages of both algorithmic and problem-oriented languages. Combining their properties, it is possible to design languages which are both easily acceptable and useful in different cases of communication between man and computer.

Application of computers in certain areas is based on the ability to isolate specific functional (and logical) wholes, as elementary tasks from which more complicated procedures can be composed. Such an elementary task within a language – or operator – can vary widely in complexity and internal structure. In this way those interested in computer use can compile operator programs as a series of such operators resulting from an analysis of complex user's needs, and are left to think at the level of these varying operators.

The approach can be illustrated through a language called PAOL-1 (Problem and Algorithmic Oriented Language –version 1).

Syntax of the PAOL-1

We will use a formalism called Backus notation for the syntax description of the language PAOL-1.

Program Definition

$$\langle \text{program} \rangle ::= \{\langle \text{program line} \rangle\}_1^{n_1}$$

$$\langle \text{program line} \rangle ::= \langle \text{label} \rangle \langle \text{operator} \rangle \langle \text{end of line} \rangle$$

$$\langle \text{label} \rangle ::= \{\langle \text{digit} \rangle\}_1^4$$

where $n_1 \leqslant n_{max}$, and n_{max} is the largest number of program lines which could be stored in the program zone.

Operator Definition

$$<\text{operator}>\ ::=\ <\text{operator's name}>\Big\{\{\,:\mid<\text{blank}>\}\ <\text{parameter list}>\Big\}_0^1$$

$$<\text{operator's name}>\ ::=\ <\text{letter}>\{<\text{letter}>\mid<\text{digit}>\}_0^5$$

$$<\text{parameter list}>\ ::=\ <\text{parameter}>\{,<\text{parameter}>\}_0^9$$

Parameter Definition

$$<\text{parameter}>\ ::=\ <\text{constant}>\mid<\text{variable}>$$

Constant Definition

$$<\text{constant}>\ ::=\ <\text{number}>\mid<\text{character string}>\mid<\text{logical constant}>$$

$$<\text{number}>\ ::=\ <\text{integer number}>\mid<\text{real number}>$$

$$<\text{integer number}>\ ::=\ \{<\text{sign}>\}_0^1\{<\text{digit}>\}_1^4$$

$$<\text{real number}>\ ::=\ \{<\text{sign}>\}_0^1\{<\text{digit}>\}_1^k\cdot\{<\text{digit}>\}_0^{7-k}\ \{E\{<\text{sign}>\}_0^1\{<\text{digit}>\}_1^2\}_0^1$$

$$<\text{character string}>\ ::=\ \text{“}\{<\text{graphic symbol}>\mid<\text{blank}>\}_1^{12}\text{”}$$

$$<\text{logical constant}>\ ::=\ .\text{TRUE}.\mid.\text{FALSE}.$$

where k is an integer number so that the total numbers of a digit in integer and fraction parts of the real number must be less than or equal to 7.

Variable Definition

$$<\text{variable}>\ ::=\ <\text{letter}>\{<\text{letter}>\mid<\text{digit}>\}_0^5$$

Symbol Definition

$$<\text{symbol}>\ ::=\ <\text{graphic symbol}>\mid<\text{non-graphic symbol}>$$

$$<\text{graphic symbol}>\ ::=\ <\text{digit}>\mid<\text{letter}>\mid<\text{sign}>\ <\text{special symbol}>$$

$$<\text{non-graphic symbol}>\ ::=\ <\text{blank}>\mid<\text{end of line}>$$

<digit> ::= 0 | 1 | 2 | 3 | 4 | 5 | 6 | 7 | 8 | 9

<letter> ::= A | B | C | D | E | F | G | H | I | J | K | L | M | N | O | P | Q | R |
S | T | U | V | W | X | Y | Z |

<sign> ::= + | –

<special symbol> ::= “ | (|) | – | . | , | : | ; | ? | !

The value of the metavariable < blank > is the empty space the size of which corresponds to a graphic symbol. The value of the metavariable < end of line > is a command symbol which defines the end of a line.

Language Implementation

There are two modes in an operator execution: program and operator mode. In program mode the program is executed from the program zone in the memory. In operator mode the last operator taken from the input, but without a label, is executed. The algorithm for the language implementation is shown in Figure 6.1.

Figure 6.1

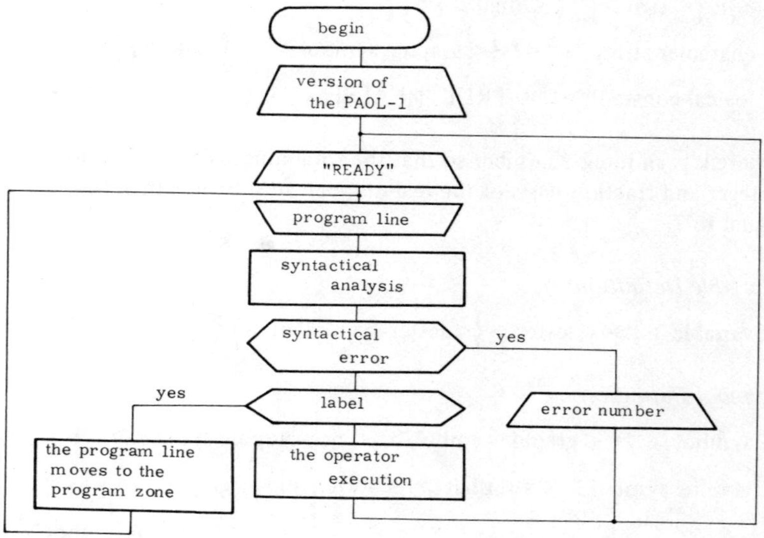

At the beginning the computer prints the report about the language version, then the word "READY", and waits for a user intervention. The user enters a program line. The syntactical analysis discovers the syntactical classes and syntactical errors. If there is a syntactical error, then the computer prints the error number by which the user can recognise the kind of the error. If there is not a syntactical error, then the computer tests whether there is a label in the program line. If there is the label, the program line moves to the program zone. If there is no label, the operator should be executed. After the operator is executed, the computer prints the word "READY" and waits for the next intervention of the user.

There is a definition table of the operators in the memory. In this table, there is information about the operator's name and the parameter list. This information is used during lexical and syntactical analysis and during an operation execution, and is entered in the table when the user defines a new operator.

Classification of Operators

Semantically operators can be divided into several groups. First, they split into two categories:

< operator > :: = < general operator > | < problem-oriented operator >

General operators are common for most applications, while problem-oriented operators characterise the particular application.

The general operators can also be divided into two groups:

< general operator > :: = < basic operator > | < structural operator >

Basic Operators

Basic operators define basic procedures for program evaluation and manipulation with program and operators. They can be classified as:

< basic operator > :: = < initialisation of zones > | <program starting> |
< adoption of new operator > | < erasing operator > |
< creating of program file > | < calling program file > |
< listing program zone > | < listing definition table > |
< input operator > | < print operator > |

Initialisation of Zones

Syntax

< initialisation of zones > :: = INIT

Semantics

This operator, which should be the first, initiates zones of constants, variable and program.

Program Start

Syntax

< program start > :: = PROG

Semantics

Initialisation of program execution. The operator with the label of the lowest number is executed first.

Adoption of New Operator

Syntax

< adoption of new operator > :: = ANO

Semantics

The purpose of this operator is the creation of a new operator. The syntax of the new operator is inserted in interactive mode through the keyboard of the terminal into the table of operators. This operator is essential for creating different versions of operator languages, to be present in the table of operators.

Procedure for Input of a New Operator

Note: Underlined text is printed by operator ANO.

ADOPTION OF NEW OPERATOR INTO PAOL-1
NAME OF OPERATOR: < operator name > < end of line >
NUMBER OF PARAMETERS
  INTEGER: < integer number > < end of line>
  REAL: < integer number > < end of line >
  STRING: < integer number > < end of line >
  LOGICAL: < integer number > < end of line >

<u>IMPLEMENTATION OF OPERATOR</u>: < implementation >
< end of line >
<u>MEDIUM</u>: < medium > < end of line >
<u>IN FUTURE YOU CAN USE OPERATOR WITH NAME</u>:
< operator name >

where are

< implementation > :: = S $\{\text{OFTWARE}\}_0^1$ | H $\{\text{ARDWARE}\}_0^1$

< medium > :: = DISK O | DISK 1 | I $\{\text{NTERNAL MEMORY}\}_0^1$

Example of adoption of operator SIGNAL:

```
OPERATING SYSTEM: RT-11       LANGUAGE: PAOL1       DATE:03-JUN-78

READY

ANO
ADOPTION OF NEW OPERATOR INTO PAOL-1
************************************

NAME OF OPERATOR: SIGNAL
NUMBER OF PARAMETERS

        INTEGER: 4

        REAL: 0

        STRING: 1

        LOGICAL: 0

IMPLEMENTATION OF OPERATOR: SOFTWARE

MEDIUM: DISK 0

IN FUTURE YOU CAN USE OPERATOR WITH NAME SIGNAL
****************************************
```

Erasing an Operator

Syntax

< erasing operator > :: = ERAS

Semantics

This operator erases old operators from the table of operators, in interactive mode.

Creating of Program File

Syntax

< creating of program file > :: MEMO

Semantics

This operator creates program file on disk 0 or 1 in interactive mode.

Calling Program File

Syntax

< calling program file > :: = CALL

Semantics

Transfers program file to program zone in interactive mode.

Listing Program Zone

Syntax

< listing program zone > :: = LISTPZ

Semantics

Prints contents of the program zone on the terminal.

Listing Definition Table

Syntax

< listing definition table > :: = LISTDT

Semantics

Prints contents of the definition table on the terminal.

Input Operator

Syntax

< input operator > :: = INPUT < variable >

Semantics

Assigns data from the terminal keyboard to the specified variable.

Print Operator

Syntax

< print operator > :: = PRINT < variable >

Semantics

Prints value of the specified variable on the terminal.

Structural Operators

Structural operators influence the formation of different algorithm structures of programs. There are four kinds of structural operators:

< structural operator > :: = < unconditional transfer > |
< conditional transfer > | < repeat operator > | < stop operator >

Unconditional Transfer

Syntax

< unconditional transfer > :: = GO TO < label >

Semantics

Transfers program control to program line number.

Conditional Transfer

Syntax

< conditional transfer > :: = IFP < variable >, < label >

Semantics

If the value of the specified variable is positive, program control is transferred to the program line number indicated, otherwise control is transferred to the operator in the next program line.

Repeat Operator

Syntax

< repeat operator > :: = REPEAT < variable >, < label >

Semantics

If the value of the specified variable is greater than zero, the value of the variable is reduced by one and control passes to the line number specified. If the value of the specified variable is equal to or less than zero, control passes to the next operator in sequence.

Stop Operator

Syntax

< stop operator > :: = STOP

Semantics

The stop operator causes termination of program execution and on the terminal is printed the message:

STOP AT < label >

Problem-oriented Operators

As mentioned at the beginning of this section, a problem-oriented operator characterises a particular class of applications. As an example of its application, a typical case in neurophysiology is now given.

All problem-oriented operators could be divided into two classes: with a parameter list and without a parameter list:

< problem-oriented operator > :: = < name > < parameter list > $^{1}_{0}$

For problem-oriented operators without a parameter list, a common parameter zone exists, whose elements are changed during the execution of operators, or else they are up-dated with a special operator DEFZON before the execution of other operators.

Examples of Problem-oriented Operators in Neurophysiology

Problem-oriented operators in neurophysiology could be divided, according to their action, into several groups:

< problem-oriented operator in neurophysiology > :: =
< data acquisition operator > | < processing operator > |
< fitting operator > | < operator-model > |
< experiment control operator > | < display operator >

Here, some operators are described as examples, while others are defined in the description of a neurophysiological experiment (see p.113 et seq.).

Data Acquisition Operators

These operators collect data emerging from neurophysiological experiments. Neurophysiological data can be continuous, like EEG and EMG signals, or discrete, like spikes from a single cell unit. Accordingly, there are two kinds of data acquisition operators for neurophysiology:

< data acquisition operator > :: = < continuous data operator > |
< discrete data operator >

An Example of Continuous Data Operator

There are two versions of this operator – with parameter list and without parameter list.

Syntax

< continuous data operator > :: SIGNAL {< parameter list >}$^1_0$
< parameter list > :: = < sampling frequency (samples/sec)>,
< duration of conversion (sec) >, < first channel >,
< number of channels >,
< common name for data files of channels >, {< medium >}$^1_0$

Semantics

This operator performs A/D conversion of the continuous signal from specified channels with a specified sampling frequency, for a specified duration, and forms files on a specified medium (if the medium is not indicated, the default medium is system disk 0). If a parameter list is not explicitly defined, the values for parameters are inserted in the common parameter zone in interactive mode through the keyboard of the terminal during operator execution.

Example

SIGNAL 256, 120, 0, 5, "EEGCH"
sampling frequency is 256 samples/sec
duration of conversion is 120 sec.
first channel is zero,
total number of channels is 5,
common name is EEGCH

As the medium is not explicitly defined, it is by default DISK 0, so after operator execution on it, five files are formed, each having 256 X 120 = 30720 data items, and the following report is printed:

```
MEDIUM: DISK 0    CHANNEL    FILE NAME    NUMBER OF SAMPLES
-------------------------------------------------------------
                     0        EEGCH0           30720
                     1        EEGCH1           30720
                     2        EEGCH2           30720
                     3        EEGCH3           30720
                     4        EEGCH4           30720
-------------------------------------------------------------
```

Preprocessing Operators

These operators perform the previous treatment and reduction of rough experimental data gathered from neurophysiological experiments. Matching the two types of data, continuous and discrete, these operators can be classified in two groups:

< preprocessing operator > :: = < continuous preprocessing operator > |
< discrete preprocessing operator >

Continuous preprocessing operators split into two groups:

< continuous preprocessing operator > :: =
< preprocessing in frequency domain > |
< preprocessing in time domain >

As an example of an operator for preprocessing continuous data in frequency domain could serve standard analysis of EEG signals. There are two versions of this operator – with parameter list and without parameter list:

Syntax

< standard analysis of EEG signal > :: =
SEEG $\{$ < parameter list > $\}_0^1$

< parameter list > :: = < sampling frequency>, < first channel >,
< number of channels >, < number of points in an epoch >,
< common name for data files of channels > $\{$, < medium > $\}_0^1$

Semantics

Standard analysis of EEG signals comprises power frequency spectrum for standard frequency bands: theta (0 – 4Hz), delta (4 – 8Hz), alpha (8 – 12Hz) and beta (> 12Hz). Every channel-file is divided into epochs and for every epoch, using a Fast Fourier Transform algorithm, relative powers of these frequency bands are evaluated, as well as dominant frequencies and their relative powers in each band. After evaluations an appropriate report is printed on the user terminal.

Example

SEEG 256, 120, 0, 5, 1024, "EEGCH"

(All parameters are the same as in the example of operator SIGNAL described above except for the fifth parameter, which means that in an epoch there are 1,024 points). Five files are analysed (EEGCH0 – EEGCH4); each file contains 256 × 120/1024 = 30 epochs, and each epoch lasts 4 sec. A part of the report for the first epoch of file EEGCH0 looks like this:

```
     S T A N D A R D   A N A L Y S I S   O F   E E G   S I G N A L S

FILE: EEGCH0    NUMBER OF EPOCH: 30    DURATION OF EACH EPOCH:4.00 SEC
------------------------------------------------------------------------
EPOCH:  1        TOTAL         DELTA         THETA         ALPHA      BETA
------------------------------------------------------------------------
POWER(%)        100.00         72.83          8.43          7.65     11.09
DOM. FREQ. (HZ)   0.25          0.25          6.50         11.00     18.25
POW. FREQ. (%)   22.92         22.92          1.45          1.10      0.48
------------------------------------------------------------------------
EPOCH:  2        TOTAL         DELTA         THETA         ALPHA      BETA
------------------------------------------------------------------------
```

An Example of a Neurophysiological Experiment

Main Idea and Goals

As an example of an experiment, where a computer system could be involved, we shall describe the attachment of the computer on-line to a conditioned reflex preparation in cats.

The very conditioning of the cat is performed by computer: the lack of leg movement, following the specific sound signal of given duration, is punished through an electric shock. As a sign of leg movement, an EMG signal is used.

It is important to note that the presentation of the conditional signal (sound) to the animal is distributed stochastically in time, during this phase. In this way, eventual rhythm-conditioning is disentangled from the looked-for effect, leaving the adaptation to rhythm for a later stage in the etological game. Imposing rhythm through a conditional signal alone in later stages, one can attain a 'clean' development of expectation (i.e. an expectation of the conditional signal developed without use of shocks).

The strategies of computer control of the cat's behaviour are based on the possibility of conditional signal removal, i.e. the signal in series which is expected, is omitted. For this purpose it is essential to know some characteristic states of the cat, and their change (eventual) with signal expectancy. In other words, the state of expectation of the conditional stimuli should be defined, and detectable on-line. To attain these goals, some experiments were performed, utilising problem-oriented operators specially designed.

Computer Training of Cats

The goal of these experiments was the computer training of cats, to form conditional reflexes on different stimuli (sound, light, etc.). A specially designed operator (named CTC – Computer Training Cat) performs not only control of the experiment, but also some evaluations and statistics on given responses. Particular attention was paid to evade rhythm, which an experimentator inadvertently produces during manual training. Randomness of inter-trial intervals is achieved, adding an increment of some time interval multiplied by a random integer number of chosen range to some base time. This operator has no parameter list, so values of parameters are inserted from the keyboard of the terminal in interactive mode in the following manner:

```
              PROGRAM FOR AUTOMATIC TRAINING CAT BY COMPUTER

INPUT FOLLOWING VALUES:

CONDITIONAL STIMULUS CS:SOUND

NUMBER OF TRIALS N:15

DURATION OF CONDITIONAL STIMULUS(SEC) TS:6.0

MAXIMUM DURATION OF RESPONSE(SEC) TR:5.0

DURATION OF PUNISHMENT(SEC) TP:0.2
```

```
PARAMETERS FOR RANDOM INTERVALS BETWEEN TRIALS:

      BASE TIME(SEC) TB:40.

      INCREMENT TIME(SEC) TI:0.

      RANGE OF RANDOM NUMBERS NR:0
```

During the experiment, values of response-intervals, and intervals between trials, are printed on the terminal. At the end of the experiment, the number of responses and failures, as well as the mean value of the response-intervals are printed on the computer terminal:

```
TRIAL     RESPONCE TIME(SEC)            NEXT TRIAL(SEC)

  1            *****                      40 SEC
  2             0.76                      40 SEC
  3             1.14                      40 SEC
  4             2.06                      40 SEC
  5             2.09                      40 SEC
  6             3.45                      40 SEC
  7             3.58                      40 SEC
  8            *****                      40 SEC
  9             0.48                      40 SEC
 10             1.18                      40 SEC
 11             2.24                      40 SEC
 12             3.01                      40 SEC
 13             3.32                      40 SEC
 14             4.47                      40 SEC
 15            *****                      40 SEC

                REPORT OF TRAINING CAT

 TOTAL NUMBER OF TRIALS: 15
 NUMBER OF SUCCESSFUL RESPONSES: 12
 MEDIUM RESPONCE TIME:  2.31 SEC
 NUMBER OF UNSUCCESSFUL RESPONCES:  3
```

Analysis of EMG and EEG Signals in the Course of the Experiment

The goal of this procedure was to show the significance of differences in EMG and EEG signals during the organisation of the conditional responses. The differences shown suggest the possibility of predicting the behaviour of the animal. In the case of computerised training the beginning of sampling of these signals was chosen sometime before the occurrence of the conditional stimulus (sound) up to the occurrence of the unconditional stimulus (electrical shock), see Figure 6.2.

Figure 6.2 TS – time of sampling; TC – time of conditional stimulus; TA – time of analysis

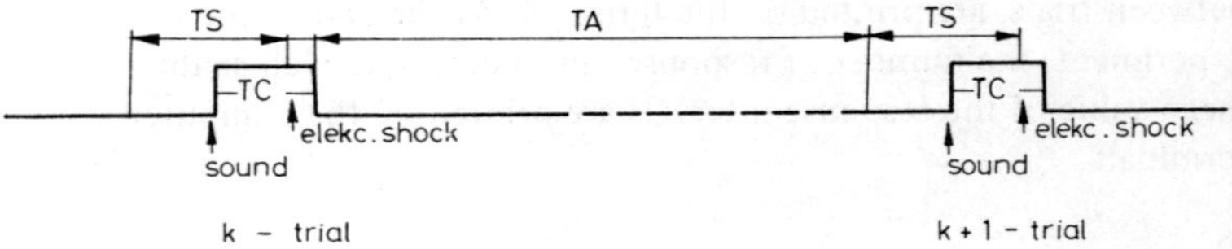

Analyses of EMG and EEG signals are performed in the time (TA) between trials and an appropriate report is given. The flow chart of this experiment is shown in Figure 6.3.

Analysis of EMG Signal

The amplitude of the EMG signal indicates the tonic and clonic states of the muscle of the active leg. So, if the parameters which define amplitude thresholds for tonic and clonic states are given it is possible to detect those states, comparing the amplitudes of EMG signals with given thresholds.

Analysis of EEG Signals

Several EEG signals were chosen for sampling. Analysis of these signals comprised a power frequency spectrum for the given duration of the epoch and parameters for the frequency bands.

Experimental Procedure

For the implementation of this experimental procedure on the computer, several operators were designed. Having no explicit parameter lists, these are inserted through the operator DEFZON into the common parameter zone.

Figure 6.3

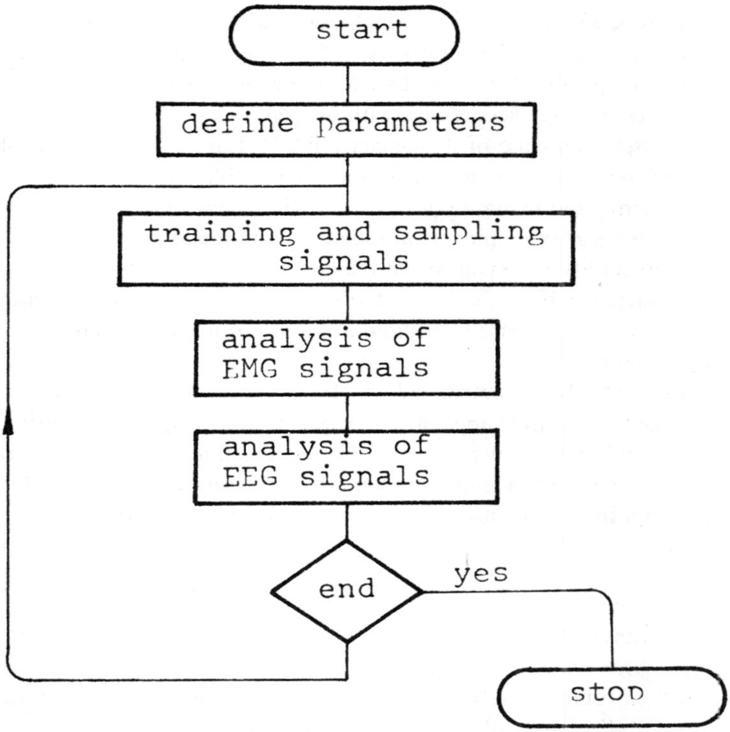

The operator program which describes the algorithm of Figure 6.3 can be represented as follows:

```
10 DEFZON       - define common parameter zone
20 INPUT N      - input number of friols
30 CTCS         - computer training cat and sampling signals
40 ANEMG        - analysis of EMG signal
50 ANEEG        - analysis of EEG signals
60 REPEAT N,30
70 STOP
```

Conclusion

The language PAOL-1 combines some features of problem-oriented and procedure-oriented languages. It is supposed to be appropriate for different users, who are not skilled mathematically and do not have computing experience. Characteristics of this language include a simple syntax and the possibility for the user to define new operators and expand the language in accordance with his needs.

In this discussion a use of the language PAOL-1 in neurophysiological experiments was chosen as an illustration. The problem-oriented operators mentioned here do not cover the whole area of neurophysiological experiments, although, as mentioned previously, there is a possibility of expansion of the language using the operator ANO. For instance, if one adds operators for modelling and fitting, the same set of operators could be utilised in neighbouring areas such as neurocybernetics.

Some ideas are being considered which could be implemented in the language for real-time application. These include assigning execution of several simultaneous operators and the implementation of interaction of experimenters with on-going programs. In addition, the elements of the parameter lists could also be arrays, their elements, and even operators.

6.5 A Hierarchy of Interactions

However the problem is solved it is essential to consider practical details of implementation particularly the limited resource in terms of computer power often available in laboratories. In addition it may be helpful to define quite carefully various levels of sophistication of users of computer software. For example, leading from the various uses described in earlier chapters it is possible to describe levels of computer use for students and research workers.

Use of complete programs, for example, for simulation in computer-assisted learning or simple statistical analysis. The computer language will be transparent to students as they respond to a dialogue written in a human language in order to set or change parameter values or supply experimental data.

Choice of models which already exist in a program is little different to *use* and the choice may be made through a human language dialogue.

Study of a model contained in a program does require that the model is expressed in easily readable and recognisable terms and is easy to find amongst other controlling statements.

Amendments to an existing model require similar ease in reading and expressing relationships.

Definition of relationships to build a program in an exploration of the behaviour of chosen functions. This is part of what could be termed computer-based experimental mathematics and would need simple graphics display features.

Creation of models requires the highest level of software sophistication demanding the logical structuring of the operations required. There are perhaps three types of models which should be considered though this is not a unique classification: (1) models from hypotheses expressed in algebraic terms; (2) models from fitting experimental data; (3) models from hypotheses expressed as system diagrams.

The ideal of an interpretive language system capable of internal redefinition is an important goal and could also allow each level described here to be self-consistent and downwardly compatible. To ensure easy portability it is likely that the system would be written in a popular high level language as this also allows a redefinition of features of the language by relatively inexperienced computer users.

PARTICIPANTS

Z. Damjanović Director, Centre for Multi-disciplinary Studies, University of Belgrade.

V.B. Bosnjaković Senior Researcher, Institute for Nuclear Medicine, School of Medicine, University of Belgrade.

B. Janković Head of Computer Laboratory, Centre for Multi-disciplinary Studies, University of Belgrade.

N. Parezanović Associated Professor of Computer Science, Department of Natural and Mathematical Sciences, University of Belgrade.

V. Spasić Postgraduate student, Centre for Multi-disciplinary Studies, University of Belgrade.

B.V. Gnedenko Professor of Mathematics, University of Moscow.

R. Lewis Reader in Computer Assisted Education, Chelsea College, University of London.

G. Franglen Director of Multi-discipline Laboratories, St George's Hospital Medical School, University of London.

R.D. Harding Lecturer in Mathematics, Department of Applied Mathematics and Theoretical Physics, University of Cambridge.

R.I. Kitney Lecturer in Bioengineering, Chelsea College, University of London.

F. Oliveira-Pinto Lecturer in Mathematics, Chelsea College, University of London.

A.W. Wood Research Fellow, Chelsea College, University of London.

One of the papers was co-authored by B.W. Conolly, Professor of Operational Research at Chelsea College. The symposium was also attended by Marija Cirić (University of Belgrade) and Hilary Suffern (British Council, Belgrade).

INDEX

analog to digital conversion (ADC or A/D) 28, 29, 39, 80, 83
analysis: Fourier 71; multivariate 27; signal 77; stochastic 64-7

cardiac 30-1, 33; *see also* intracardiac shunt and radionuclide angiocardiography
Central Limit Theorem 67
clinical (clinician) 16-19, 26-39
computational: aid 13; analysis 69-76
computer: interface 28, 39; languages 94-103, 104; assisted learning 15, 67, 84-93, 120
continuous 11, 57, 78-9
Control and Simulation Language (CSL) 101, **102**
control systems 41, 43

data: hard 17, 18, 26; soft 17, 18, 26; analysis 14. 26
deterministic 12, 56-67
diagnosis (diagnostic) 17-22, 26-39
discrete 11, 57, 100
domain: amplitude 80; frequency 80-2; time 80-3, 100

electrocardiography (ECG) 26, 35, 43, 78
entrainment: 41-55; stable 46; metastable 46, **47-55**; unstable 47, **51-2**
equation: difference 56-66; differential 54, 56-67, 69, 96
event: 100, 102; point 78

FORTRAN 10, 81, 84, 95, 99

Fourier: analysis 71; analyser 13

gamma scintillation camera (GSC) 28-39
General Purpose Simulation Systems (GPSS) 10, 99, **101**
graphical 10, 69-72, 85-92

heartrate variability 41-55

imaging: blood pool **35-6**; infarct **36**; myocardial **36-8**
intracardiac shunt **32-5**

languages: programming 10, 94-8, 104-15; simulation 10, **98-103**
logistic law 61-2

multi-disciplinary 9, 11, 26, 27, 84

operators 10, 104-20

probabilistic 56-67

radionuclide cardioangiography 28-39

SIMSCRIPT 101, **102**
SIMULA 101, **102**
simulation 10, 11, 14, 84-8
spectra: amplitude 80-3; frequency 80-3; phase 80-3; power 42-55, 118
stimulus: 78-84, 116; thermal 41
stochastic 57-67
student 14, 15, 24, 77-8, 86

Transform: Fourier 74, 82-3, *see also* analysis, Fourier; Laplace 82